This Chosen Place

Other Works by Max Evans:

Southwest Wind
Long John Dunn of Taos
The Rounders
The Hi Lo Country
The One Eyed Sky
The Mountain of Gold
Shadow of Thunder
My Pardner
Sam Peckinpah—Master of Violence
Bobby Jack Smith: You Dirty Coward
The Great Wedding
The White Shadow
Xavier's Folly and Other Stories
Super Bull—and Other True Escapades
Rounders Three (a trilogy)
Bluefeather Fellini
Bluefeather Fellini in the Sacred Realm
Spinning Sun, Grinning Moon
Broken Bones and Broken Hearts
This Chosen Place

This Chosen Place

FINDING SHANGRI-LA ON THE 4UR

Max Evans

Foreword by Tom Lea

Published by the University Press of Colorado
P.O. Box 849
Niwot, Colorado 80544
(303) 530-5337

The University Press of Colorado is a cooperative publishing enterprise supported, in part, by Adams State College, Colorado State University, Fort Lewis College, Mesa State College, Metropolitan State College of Denver, University of Colorado, University of Northern Colorado, University of Southern Colorado, and Western State College of Colorado.

The paper used in this publication meets the minimum requirements of the American National Standard for Information Sciences—Permanence of Paper for Printed Library Materials. ANSI Z39.48-1984.

Library of Congress Cataloging-in-Publication Data

(to come)

10 9 8 7 6 5 4 3 2 1

Title page photo: Stylish women of 1883 fishing on a portion of the Rio Grande that is now part of the 4UR. *(Courtesy, the American Museum of Fly Fishing, Manchester, Vermont.)*

A Dedication of Memories

To:

Herman Eubank, my last living war buddy, who fought for the boxing championship even though I was his manager, and who, like General John C. Fremont—though well over a hundred years later—nearly froze to death in the mountains above Creede, Colorado. Every day, while I scribble my life away, he has enough good sense to go fishing.

To:

Robert Conley, whose writings of the Cherokee *Real People* are far too slowly receiving their justified worldly acclaim. Even while the *little people* haunt him into recording the beautiful and historic worlds of the Southeast forests, an occasional Wild Turkey keeps him smiling.

To:

Ed Honeck, one of the original *Orange County Cowboys*, who has hung tough, against unbelievable odds, longer than anyone I know—and with great good humor.

To:

J. Lee Smith, deceased, my old high school principal who somehow had the vision to work it so I could read the old masters like Balzac, Dostoyevsky, and Cervantes, while other students were doing standard homework. He never missed a wing-shot, either—when we were hungry.

All of these people have, or had, that rarest combination of human traits, dead-on loyalty to earned friends, knowing it is not to be qualified. We shared thousands of laughs at how ridiculous we all are, especially ourselves.

Contents

Foreword   xi

Acknowledgments   xiii

Introduction   xv

1. This Chosen Place   1
2. The Valley of Sudden Death and Bountiful Giving   9
3. Roots of Home and Guts Aplenty   12
4. Back When the Horses Came Running   23
5. The Launching of a Giant   30
6. Consummate Courage, White Death and Making History   42
7. The Days of Surging Power and Overcoming   57
8. General Palmer: Civil War Hero and Owner of the 4UR   64
9. The Phipps Clan of Colorado: Doers and Givers Supreme   75
10. Explosions of Energy Catapulting Through Canyons of Dreams   85
11. There Is No Night in Creede, or So They Once Said   98
12. Go West, Young General, Go West to the Colorado Rockies   111
13. Charles H. Leavell's Power on the Ground and in the Air   125
14. The Leavell Company Joins the Cold War with Hot Projects   141
15. The Wiles of Walton and the Last Trail Winding   148
16. Buying Up the World and Hoping for Paradise   157

CONTENTS

17. The High and the Low: A Necessary Love Affair   175
18. Natural Living and Some Tranquillity   184
19. Repairing Paradise Just a Little   190
20. Time of Rebirth and Friendships Rare   204
21. Seeking the Lost Lakes: The Demolishing of Burdens 217
22. The Daily Workings: Fulfilled Yearnings   222
23. Mines and Moguls: Saviors Come Calling   235
24. Genes of History: The Family of Perpetuation   238
25. The Big Bow Knot and the Forever Time   255

Sources   263
Index   271

Foreword

Charles Leavell has been my close friend for half a century. Through all those years, his wife Shirley and my wife Sarah have been as strong in their friendship as Charles and I have been in ours. The four of us, in a long time of living, have drawn together in deep and lasting—almost family-like—affection.

Charles is a self-made, well-made man. He was born in El Paso on the 26th of February, 1911, son of the late Charles Leavell, Sr. and Mabel Walton Leavell. Both his parents were fine citizens of El Paso. In a prosperous home on Federal Street they brought up a close and affectionate family of four: Charles and his three sisters, Imogene, Kate, and Josephine.

In childhood Charles fought a harsh battle and won a noble victory over a crippling opponent, tuberculosis of the bone. That victory left him with a game leg, but it built into him a power of will, a confidence in himself and his own hardihood—a conviction lasting to this day--that enough resilience to misfortune can convert adversity to advantage.

In the space allotted to me here, it's not possible to sketch even roughly the story of Charles Leavell's professional performance as the head of the firm he created in modest circumstances, and then guided and directed to great success, in a highly competitive field. The company spread its activity from a local enterprise centered in El Paso to an organization—still centered in and directed from El Paso—performing work worldwide, sometimes alone as sole contractor, more

·From an introduction by artist/writer Tom Lea at the El Paso County Historical Society Hall of Honor Dinner, November 22, 1991.

often as a joint venturer with construction giants such as the Utah Company, Morrison-Knudson, Peter Kiewit Sons, and many others.

The structures built by C.H. Leavell and Co. are very, very many, their purposes as diverse as their locations, from Milwaukee, Wisconsin, to Monrovia, Liberia: bridges, skyscrapers, missile sites, banks, dams, river locks, hotels, nuclear power plants, port wharves, schools, NASA test facilities, hospitals, telescope domes, water treatment plants, churches, shopping centers, post offices . . . just name it, Charles Leavell has probably built one! To see the sort of thing he's built in El Paso, look at the State National Bank Building downtown, the First Baptist Church out on Montana Street, the education and Engineering building over on the UTEP campus.

I think often about what must go on inside my friend Charles's skull, and the satisfaction he must feel as a creator remembering shapes of structures, raised by his own know-how, to stand stout and useful and proud against the sky.

How would I describe the outstanding character and personality of Charles Leavell? I would say he is a man of great energy and generosity, tough-minded in defense of his beliefs, yet fair-minded too, with an unfailing sensitivity and sympathy for others. He listens. Possessed of a strong sense of humor, he is familiar with irony without bitterness. He has a keen sense of adventure and the enjoyment of it.

Charles and I have traveled many trails together, afoot and horseback. We've sailed together in little boats and big ships. We've ridden together in pickup trucks and limos, taxicabs and jeeps. We've been together in a lot of places.

We've shared exalted moments in the snowy tops of the Wind River Mountains. We've listened to the windsong in the rigging sailing the "wine dark sea" of Homer's *Iliad*. We've ridden with gauchos on the pampas of the Argentine. We've even been around to lean—both at the same time—on the Leaning Tower of Pisa!

My friends, I know the man well and I present to you with great delight a great El Pasoan—Charles H. Leavell!

Tom Lea

Acknowledgments

I give great gratitude to Shirley and Charles H. Leavell and their family—son, Pete, and daughter, Mary Lee—for their unhesitating cooperation and technical help in weaving the strange history of the 4UR realm into a book. Among many other courtesies, all of the photographs in this book, unless otherwise indicated, are from the Charles Leavell collection.

To Marc Simmons, one of New Mexico's finest historians, I give heartfelt thanks for his recommendations and his help in gathering material on Zebulon Pike. The effort of our friend, Slim Randles, New Mexico's most unheralded writer, must be noted in clarifying the real role of General Palmer in the region. He delivered as true as a mountain-trained pack mule. To noted historians, Dr. Richard Ellis of Fort Lewis College, Durango, Colorado, and Dr. Thomas J. Noel of the University of Colorado, Denver, my eternal thanks for your prompt readings and incisive criticisms to help properly guide this work to fruition. Jim Hemesath of Adams State College and David Goetman, of that school's library, were courteous and generous with their time. Our thanks. To rancher Jimmy Bason, and authors and historians Alden Narajano, Ruth Marie Coleville, Phil Carson, Bill Lynde, Ruth Armstrong, Rhoda Davis Wilcox, Harvey L. Carter, George LaVerne Anderson, Edwin Lewis Bennet, William A. Settle, Jr., James D. Horan, Leland Feitz, Toby Smith, Marshall Sprague, John S. Fisher, George Foster Peabody, R. R. Bowlser, John J. Lipsey, Virginia Chappell, Virginia McConnell Simmons, Luther E. Bean, Beryle Vance, Andre Pagett, S. L. Korzeb, Rebecca Craver, Mary Lee Spence, Adair Margo, Kathleen G.H. Jerter, Anne Dingus, Mary Lewis Kleberg, Keith C. Russell, Izak Walton, and Bradford Young, my gracious thanks for your written words that helped so much in painting the word pictures of the 4UR realm. If I have neglected anyone it is because of my worn memory, not my heart, and I ask forgiveness.

Then, of course, we must pay homage to the people who have scattered their skills and souls along Goose Creek: Kristen and Rock Swensen, half-century guest Ed Cook, Bill Geis, Ed Wintz, Katus Walton, and Toni Shaw.

Compliments and deep thanks to Dick Gober for giving of his own crowded time to expertly do the 4UR realm map.

Abrazos to Tom Lea for his foreword and for being a great artist and friend.

And always to Pat, my wife, who helps in uncountable ways and is doing the book jacket as well.

Lastly, to our dogs, Foxy and Shadow, who give love even when you're telling the truth or being curmudgeonly—even though both actions are the same thing nowadays.

Introduction

When I was first approached about doing this book, I was extremely wary of getting involved. I had done a number of essays and other nonfiction works—including a book, published long ago, on outlaw-gambler-entrepreneur, Long John Dunn of Taos—but at first, I wasn't exactly sure how to make this one work.

When I finished what I considered my major work to date—the *Bluefeather Fellini* and *Bluefeather Fellini in the Sacred Realm* novels—I thought I had faced and conquered almost every kind of writing challenge possible. I had spent over thirty years of living and researching—and six years of actual writing—on this duo before I finally decided I might be accomplished enough to tackle the actual writing. I almost caved in when I started going through those decades' worth of notes and written pages, some of it dating back to 1950 when I was just a kid. The realization smacked me that I had enough material for ten thousand pages, or at least twenty to forty books. I realized I'd have to condense—not eliminate—this to between one and two thousand pages. I finally did get it all in about eleven hundred pages.

Now here I was being presented a nonfiction book, not nearly as long but just as difficult, and on a lesser scale I would have to do the same thing here. At first my brain turned to reinforced cement, then slowly my initial hesitance eroded as the project started to reveal itself.

Mr. Charles H. Leavell is a very forceful and dedicated man. The book would have to be focused on him, and even more so on the "dream place" that he had pursued for so many years—the 4UR Guest Ranch on Goose Creek and the Rio Grande. The 4UR was not only a world-class trout fishing location, but also was set in the majestic mountains surrounding the once-wild and historical town of Creede, Colorado.

I realized that to show the importance of the place, and all the blood and skin shed to finally complete Mr. Leavell's dream, we would have to go all the way back to the first inhabitants of the 4UR—from the prehistoric Indians of the area up to the Ute Indians—then forward through the history of the whole 4UR realm. That history includes many notables: pioneers like de Anza, Fremont, and Pike; some of the world-famous outlaws and prostitutes who followed in the early mineral mining days; then, of course, the 4UR's first real developer, General William J. Palmer, who was followed by the wealthy and powerful Phipps family of Denver. What I found here was about five books that somehow had to be condensed into one.

Mr. Leavell and his wife, Shirley, are the kind of people one can't help falling in love with, and they had a surplus of energy to give to this project. Not only had Charles Leavell—with the love, support and dedication of Shirley—conquered a great part of the world out of El Paso, Texas, in the world of industry, but he had done so with a lame leg from his early youth. He had done it all with class and a great sense of fine things—art, writing, friends and, above all, a choice sense of humor. Together they had raised a family, Pete and Mary Lee, of equal feelings and quality. These children had parents who gave them freedom and at the same time strict regimentation and guidance. The couple's life-risking adventures in Khartoum, Liberia, China and the jungles of New Guinea and Africa were enough to manufacture a lifetime supply of adrenaline, but were also somehow turned into great fun by their attitude toward their very narrow escapes.

Now, above all this, Charles is an addicted and gifted fly fisherman, and naturally so are most of his family and friends. Therein resided the essence of his distant and difficult dream. He could easily afford to buy a host of ranches, which he did. But no matter how luxurious and profitable they were, or how much fun they afforded to family and friends, he could never get the 4UR out of his mind—his very soul. It was his total obsession.

I called the director of the University Press of Colorado, Luther Wilson, who had successfully published the Bluefeather Fellini books, and described to him how I wanted to set up an *essence* of the history of the 4UR in the very beginning of the book so the reader could understand without a great waste of words the events that led up to the present-day 4UR paradise. He responded with enthusiasm, saying that he could see no other way to present it. With a developing admiration for the vast accomplishments of Charles and the rest of the Leavell family, I became thoroughly excited at the enormity of the challenge.

So, I decided I would research the beginnings of the book the very best I could and then have the facts checked by two academic historians, whose teaching and writings I admired: Richard Ellis, western historian at Fort Lewis College in Durango, Colorado, and Tom Noel, western historian from the University of Colorado at Denver. Then it would all be up to me.

It would be no stroll in the woods, I knew. However, I had a feeling I was on the right track and it grew into a strong one. Mr. Leavell was at least as great as those prominent figures who came before him on the 4UR—more so because he overcame more than they had. It is and was, in finality, the much-maligned but true American dream. He had accomplished this with immense dedication and dignity.

The thing that struck me most, above anything else, was the fact that such empire builders as General Palmer, Allen Phipps, and the Leavell family all became more fulfilled, more content and more giving people after their contact with the 4UR. This attitude applied totally to every worker there and every guest I've met and talked with. I too, like Mr. Leavell, succumbed to the alluring enchantment of the 4UR, and all the people, past and present, I was to meet who had been associated with it, including all the denizens of the wild National Forest surrounding it. So how could I not attempt to write the many journeys by great men and women across the vast 4UR realm? How, indeed?

Another, no-less-important lesson was attached to the 4UR's story. Everyone in the world is interconnected—their souls, their shadows, their paths, moving in and out of the blending, overlapping, spirals of circles, directly or indirectly touching. The pioneers, the leaders—because they are fewer and on the point—cross each other's trails and endeavors the most and the clearest, and their deeds—good and bad—fuse into one massive entity inseparable from the wondrous earth that sustains them in spirit and flesh.

Yes, Mr. Charles H. Leavell had crossed the same wilderness paths and broken bread with giants. He is one himself: a remarkable, moving gift connected to this precious parcel of earth. Almost every day he still goes to his El Paso office, which is filled with beautiful art and efficiency and overseen by his skilled secretary, Lori Andrulis, whom he calls his "brains," to work on some project. When not there, he and Shirley may be visiting any part of the world. More likely, they will be at the 4UR, which after twenty-six years is "brand new every day."

The ranch holds in its wholeness a story of inspiration and place, a very special spot on this harassed globe. I've joyfully given all I have to this work and place from wherever the heart and soul are located, just as all those others and Charles H. Leavell and his family did before me.

This Chosen Place

CHAPTER 1

This Chosen Place

The world is truly wonderful if a person is in the right place at the right time, and there are certain locations on this earth that are special to everyone who nears them. The sacred Taos Mountain hovering over the ancient Taos pueblo is one of these places. It dominates this valley, which still holds the vibrations of medicine men and women, mountain men, great Spanish and Mexican haciendas and ranches with their vast flocks of sheep and herds of cattle and horses. This is the place where fine artists come from all over the world to breathe in and paint the green sage, the golden aspens, the blue mountains and the diamond-dust air. Others come from around the globe just to absorb the mystical magic surrounding the great mountain.

In Egypt there is a valley of the "Old Kingdom" where the pharaohs and hundreds of thousands of workers built the massive, incredible Sphinx, the pyramids of wonder and the tombs of the royals, filled with astounding art of gold and skilled wall drawings of both history and mystery—alluring, intangible, but special to all the world.

The Taj Mahal, in India, is another place of worship and haunting beauty; photographs alone have enchanted all who gaze upon them. And the world has been blessed with the vast plains of Tanganyika in Africa, near one of the birthplaces of man and a great vessel of plenty for millions of wild creatures that wander with grace and savagery, death and dignity, sound and silence—a veritable cauldron of conscious life.

There is the copper canyon of Mexico and the Grand Canyon of Arizona, where brightly hued and endlessly wind-and-water-sculpted sides plunge down toward the center of the earth. Each layer of that

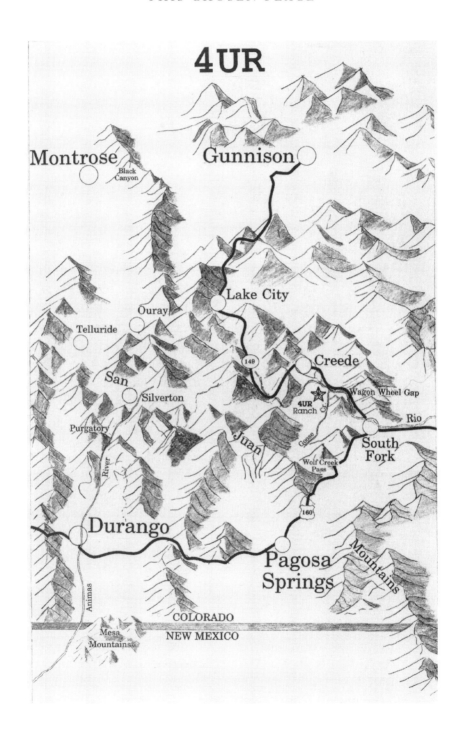

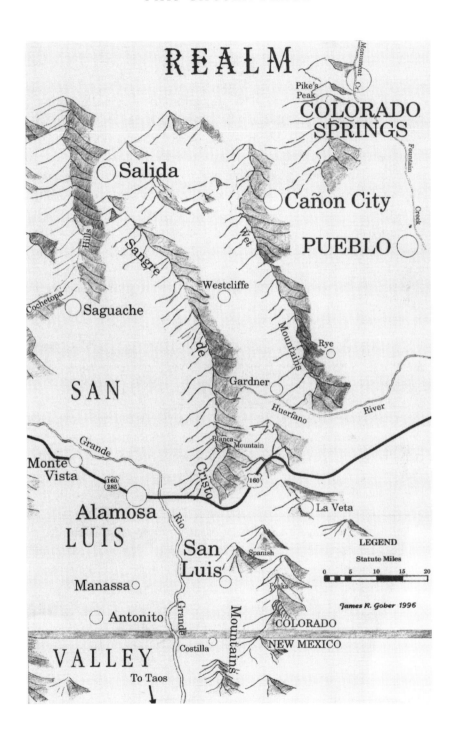

3

earth represents millions of years of existence and creates in those who view it an unfathomable feeling of awe and an unanswerable feeling of thankfulness for being a tiny portion of it all.

There is also such a place in Colorado—the 4UR Ranch. It adjoins the great Rio Grande and a main tributary called Goose Creek for eight wondrous miles. The Rio Grande's headwaters gather in liquid life above the nearby town of Creede, Colorado, coursing through the narrow and magnificent palisades of Wagon Wheel Gap and then plunging down to irrigate the farms and ranches of the over seventy-five-mile-wide San Luis Valley. It flows on past Taos, Santa Fe, the middle of Albuquerque, by Truth or Consequences and Las Cruces, New Mexico, curving at El Paso and on through the Big Bend country, giving fertility, life and sustenance to the Rio Grande Valley of southwest Texas. After struggling eighteen hundred miles, it empties into the Gulf of Mexico at Brownsville, Texas, and Matamoras, Mexico. The 4UR and Goose Creek are part of all that and more.

When Charles H. Leavell first turned the old Chrysler hardtop onto the 4UR land, he knew he had found another one of the world's special places. Still, it took many travels and adventures of body, mind and soul to fully reveal the uniqueness of the 4UR and its environs.

It was the summer of 1944. Thousands of Allied troops had died on the beaches of Normandy and were bleeding their way through the fortress hedgerows to conquer Hitler's massive machinery of slaughter and dominance. Although Charles had tried to join the military, he was turned down because of a very bad hip, acquired when he was only three years old. It was this lame leg that cranked up his natural fortitude and drive and plunged him into a life of creating many mighty things. He would enter a combat zone of his own, building things for the military. *Builder* is one of the key words to describe Charles Leavell, then and to this moment—a builder of increasingly beautiful things.

During the first two years of commitment to his first government contract, Charles had not seen his beautiful young wife, Shirley, and their two small children, Pete and Mary Lee, at all; on the third year they shared only a few scattered and stolen hours. So, for their first real vacation time together, he wanted a unique place to take them. He had heard from his close friend, Tom Lea, the great border artist/writer/

war correspondent, about the unbelievable fly fishing in a setting of lush forests full of wild game called the 4UR Guest Ranch in southern Colorado. It sounded absolutely perfect—a few miles south of Creede, Colorado, joining the originating waters of the Rio Grande. The decision was made with exuberance all around.

The excitement built as they made plans to spend an entire month together in the great outdoors. The children were almost giddy. Charles and Shirley were not far behind in their slightly more controlled elation. It really was intended to be a time to remember. They certainly did, for their entire lives. In fact, that trip would turn out fruitfully fateful for uncounted numbers of people to come.

Charles, being a Stanford graduate with a degree in electrical and construction engineering, had no trouble rapidly improvising a rig on top of the car to hold their camping and fishing equipment, plus other necessities. It was wartime and almost everything was rationed, so he went across the border from their El Paso home to Juarez and bought several five-gallon cans of gasoline. These were also lashed to the top of the car.

Driving north to Santa Fe, the Leavells stayed in the historic and beautiful hotel, La Fonda de Santa Fe, the first night. Charles's voice revealed his love for his family and this memorable time as he related, "And boy, having a vacation with my beautiful wife in that big old hotel was a dream. The kids were involved in their own worlds of wonder, as kids always are.

"A couple of days later, near Alamosa, the kids were yelling and hollering in the back seat. In my totally absorbing work for the government, I was unaccustomed to the natural antics of children, and I admit they were distracting, but I do remember turning there at Alamosa down highway 160, on to South Fork on the northern edge of the huge San Luis Valley, then turning up 149 into the mountains with the big green Rio Grande rushing down that canyon and it got boxed in more and more by the bluffs on each side. There was nothing in the canyon in those days but the railroad track, the narrow two-lane highway and that wondrous talking river. Then I remember, I will forever, seeing those awesome palisades. I was stunned."

Even Charles's highly tuned engineer's mind could not conceive of how the little beginning of the Rio Grande had ever cut its way through the heart of the Rockies—and six hundred feet of granite—so it could perpetually course on down to the Gulf of Mexico, feeding and watering all those wild animals and plants for uncountable millennia. Charles recalls, "As we neared those majestic vertical cliffs . . . boy, my old heart began beating so loudly I could hear it and I got more and more excited with each breath. They called this Wagon Wheel Gap and that's where it was really narrow and magnificent—with big ponderosa pines all around." It is unlikely that anyone will ever truly know how the Gap got its name. Some say Kit Carson's wagon broke down here and a wheel was all that was left. Others say this explorer or that explorer from Fremont to the Spanish left the wheel. No matter; a wheel was found and forever that will be the name.

"We turned off left across a bridge over the Rio Grande, past a lovely little lake, up a narrow dirt road and then there it was, the outbuildings and all right next to the swiftly flowing Goose Creek, tributary of the Rio Grande.

"The mountains dodge back higher and higher in greens and blues with the air all around clean, cool, invigorating," Charles said emphatically. "The air had a restorative feeling about it as if it were blessing you."

He was hooked, just like the rainbow trout were in Goose Creek. His heart pounded so hard his breath had to speed up to make up the oxygen. At that moment he swore he would fulfill his dream here someday. That dream had always been in his mind: a vision of a place full of wild beauty and wild creatures, with a rambling, talking creek full of active fish surrounded by great mountains and all their well-fed natural inhabitants. Yes, he knew this was it—the place where he and Shirley would retire to play, but also to work (possibly helped by their grown children) at improving the land and live water they all already loved.

If Charles had realized how long it would take for just the opportunity even to make a purchase offer on the 4UR, or how many separate worlds of contracting would be completed before then, or how many other ranches he would own, or the massiveness of the circuitous route he would travel—crossing over the trails of noted pioneers—before he could get to that special moment of immense financial cost, when he had to say yes or no and write a check for a fortune, his vision might been imbued with a far different emotion. Charles can talk about it now, though.

"It was called Wagon Wheel Ranch and was run by a fellow named Arthur Sharp and wife, Geneva—and he fit his name, let me tell you. His wife did the cooking for the outfit and Sharp disciplined the guests . . . and he did have some prominent guests. Before our arrival, [Chief] Justice Earl Warren had just left and old Lawrence Melchior, the great opera singer, was still there. He had the biggest belly I ever saw on anybody in my life, but Shirley said girth was necessary to emit such powerful sounds. If that was the case he could have sung on Goose Creek and been heard on the Nile. Ed Wintz, a young cowboy who had just been dismissed from the service, was a fine all-around hand. He was a guide, horse wrangler or anything else that had to do with enhancing the 4UR operation. He was an important part of the 4UR then and would be for all of our lives later.

"I remember pulling in front of that upper bunkhouse called the Holy Moses. I rented two rooms at only $32.50 a night, and that included food, lodging and fishing—if you could get to the stream. The roads were terrible then. We had a wonderful vacation with the good food, and horses for the kids to ride. Yes, I'd fallen in love with the place. Of course, the fishing was what solidified the lifelong romance. You never forget your first time on a special stream. Never.

"I'd worked my way to an opening in some brush where the stream flattened out a bit. With my waders and fly rod I eased out into the middle and faced upstream and then downstream. It was the most beautiful water I had ever seen at that moment, half in shadow and half in stunning sunlight. The sparkling, dancing water ran around my legs like it was talking to and caressing only me in all the world. And that's the way I felt at that moment. I was so hypnotized that I don't remember what flies I was using or how many times I had cast before I felt the eternally magic touch and tug of the trout trying to feed on the leader-connected fly. And that line connected through the rod into my hands, and straight into my heart. Oh boy, the only way you know the wonder is to do it. I landed that first trout here even if I was in a daze. When I netted it the sun enhanced all the many iridescent colors of its sides out for my tired eyes to feast upon and it gleamed like its name. Boy, a big old rainbow trout wet and shiny and fat and as beautiful as anything I'd seen since I'd first met Shirley."

Long before the Leavells, many famous, wealthy and powerful people had owned this land with its permanent, medicinal hot springs, including General William Jackson Palmer of Civil War fame, who owned the Denver and Rio Grande railroad and much else. Later there were others, such as the Phipps family of Denver, whose patriarch was a U.S. Senator from Denver, and young Gerald Phipps and Allan Phipps, who owned the Denver Broncos pro football team (one among the many parts of the widely diversified Phipps family empire).

It was not, however, in Charles Leavell's nature to think only of the rich and powerful. He respects professionals in any calling—teachers, cowboys, miners, fishing guides, chefs or whatever—and one had better live up to professional standards if one hopes to gain and treasure his friendship. If Charles Leavell has a major fault, it might be this demanding more than some are capable of giving—but considering what he himself has given and overcome, this trait must be accepted, if not totally understood.

This vacation from building the "buildings of war" unknowingly set the entire Leavell family on a journey of encounters with vast wealth, power, a growing love of nature, art and so many tragic and glorious moments of living that one can only marvel at the seemingly impossible odds of the family ever returning to Charles's dream—the 4UR Ranch. They would travel through eons of history and make some themselves on the trying journey.

CHAPTER 2

The Valley of Sudden Death and Bountiful Giving

The San Luis Valley is over a hundred miles across in some places and is more than three times the size of Delaware. It is not just this valley, but much of the southern Colorado mountains that form what is referred to here as the 4UR realm.

The valley and these mountains have always been inter-dependent, one giving to the other. A simple example is the fact that the entire area, with all its farm produce, is water-fed by various valleys that open from the Rocky Mountain streams. Besides the water, the mountains give timber, minerals and recreation. Moving the other way, the valley has supplied the mountain people with vegetables, other foodstuffs and workers for the mines, and acted as headquarters for supplies of all kinds. The two-way giving is endless.

This vast mountain/valley area—that is the concern here—is roughly cornered by Colorado Springs on the northeast; Montrose, Colorado, on the northwest; then down to the southwest corner at Aztec, New Mexico; over to Costilla, New Mexico, as the southeast corner; and returning north to Colorado Springs to complete the ragged and certainly jagged rectangle of the 4UR realm. All this territory was once Ute Indian country, but long before them it was a land of the prehistoric peoples and wild animals. Great herds of antelope and buffalo roamed across the grasses and the bush-lined waterways of the San Luis Valley. The mountains and foothills were flush with elk, deer, turkey, grouse, and uncountable other small game.

Large haystacks in a field with Mount Blanca in the background. (Courtesy, Adams State College Library)

There were vast numbers of bear, bobcat, coyote and cougar, as well as golden and bald eagles, to keep all this in balance.

Regally holding its queenly snow-capped head 14,363 feet in the pure air is Mount Blanca, near the northeast side of the valley. This huge peak dominates it all, even subjugating the Great Sand Dunes (which are now a national monument) at its western foot. Is it any wonder at all, then, that the Utes, upon first seeing and experiencing this seemingly immeasurable land, knew they had found an earthly paradise? They worshipped it and often defended it to the death. They had fish, meat, hides and beauty enough to last for several forevers. Could any other than the Great Spirit have created this place especially for them?

Easily accessible from spring to fall were the creek, medicinal hot springs and valley now known as the 4UR Ranch. It was claimed by three bands of the southern Ute tribe before the coming of the Spanish and the much later arrival of multitudes of nationalities—the so-called white people.

The hot springs were a worshipped gift coiling up from the sacred middle earth. The Mouache, Capote and Tabeguache clans hunted outward from the springs as late in the fall as they dared before packing their meat out over the great pass to Pagosa Springs.

Legend often says that the 4UR springs were called Little Medicine and Pagosa Springs were Big Medicine. This may be true, though Ute tribal historians are not certain. Nevertheless, that is what they were called according to Alden Narajano, Southern Ute tribal historian. The natural hot water and the campground on the 4UR were used in the spring by the Ute people for their special bear dances and throughout the year for relaxation, the easing of aches and pains and the healing of battle wounds. Though most historians of today do not believe it, the Utes say their ancestors sometimes camped there for entire winters, because they had harvested plentiful game earlier and found solace in the hot springs from the deep snow and freezing temperatures. There certainly would have been plenty of hides for shelter and wood for fires, it seems.

In the early 1800s, the Comanches made an attempt to take the precious valley from the Utes. The Utes fought valiantly, and probably desperately, against the rightfully feared Comanches and finally repelled them. Losses were not recorded, but they must have been heavy.

Ironically, on another occasion the Utes camped in the valley near Wagon Wheel Gap and found a group of early white settlers in a pitched battle with attacking Arapahoes. The Utes went to the settlers' aid, driving off the Arapahoes. One cannot help but wonder how their sons and daughters, and even some of the elders, felt about this when white people later removed them from this special place.

Before that, the Spanish had come, bringing with them, among the most important of the things from across the oceans and Mexico, horses and firearms. It has been no different throughout recorded time and the rest of the world—explosive powder and horses forever alter the way of life of all they touch. The Spanish incursions into the land of the Utes met with wildly mixed results, but then and thereafter, history inexorably drew animals and people to the land surrounding the natural jewel that is now the 4UR—finally to mesh there in both spirit and reality.

CHAPTER 3

Roots of Home and Guts Aplenty

Charles H. Leavell's obsession with the 4UR could partially be explained by his prior history. His family had been in the ranching business since 1896, long before the Mexican Revolution. His father, Charles H. Leavell, Sr., a tubercular from Georgetown, Texas, had gone to El Paso, Texas, and had later inherited enough money to buy a spread called the Clint Texas Ranch. It covered more than a hundred thousand acres in rough semi-desert and mesa country. "It was just old desert land on the border," Charles recalls.

Some of Charles's first memories are of his father telling stories of rustlers and Indians drifting across their ranch land, marauding the line camps and stealing Leavell cattle. Lots of trouble for all, but exciting tales for a young man of destiny born to the ranch life. To avoid all the trouble surrounding and upon the Clint Texas Ranch, Charles's father decided to sell it, but very soon he felt like a man without a soul.

Charles explains, "My father was a very rugged, wonderful, fine man, died quite young, I thought—when he was fifty-nine—but he loved ranching like I do. And he bought another ranch on the edge of the Big Bend Country of Southwest Texas, north of Sierra Blanca. It was called the Figure Two Ranch and that's where I grew up.

"It was hard, wonderful, challenging, adventurous and above all a hell of a teacher to the young. It had forty thousand acres and was on top of the Delaware Mountains, which is the great range that extends south from the Guadalupes [Texas's highest point]. And he had about five or six thousand acres down in the Salt Flats where they'd had the Salt Wars. This was an isolated war between ranching families who

The horse remuda at the Leavell Big Bend (or Figure Two) ranch, with the two wild paint mules to the right.

Old man Daugherty, who owned the ranch to the north of the Leavell ranch (left); "Blowhard," one of the Leavell ranch workers (right).

claimed the valuable mining rights to the bed of salt in a dried-up lake. It caused several killings. The festering anger, mistrust and continual violence was quieted when the Texas Rangers came in the mid-1870s and finally parceled the area and placated the warring groups. All in all that part of the land was pretty worthless country. But the mountain country was gorgeous. It was pinon and cedar, and brushy rough canyons full of deer. In the meadows and flatter pastures there were a lot of antelope and I used to go up there with my dad from our town home in El Paso."

All the ranch adventures of this boy, and all his great accomplishments for the rest of his life, must be recognized as a miracle of resolve, fully understandable only against the backdrop of historical fact. When Charles was three years old, his hip began to pain him, and he fought desperately against both the growing agony and the loss of strength in the joint. He was taken to the Mayo Clinic and there was diagnosed as having tuberculosis of the bone. They fused his hip and he did not walk without crutches again until he was ten years old. By the time he was eighteen, by sheer grit and swimming exercises, he had gained back his strength. Of course, part of that "will" did come from his father. Right after the operation Charles's father gathered the family around—Charles's mother and his three sisters, as well as Charles—and laid down the law, as ranchers were known to do then. (They probably still do, but maybe to a lesser degree now.)

His father said, "Now we all know about Charles's operation, but from now on there will never be another single word about this injury uttered in this house. There is no handicap here. It does not exist. He is a normal, healthy boy and will live a normal, healthy life." And that was the way it was, and rightly or wrongly it worked magically well.

Charles talks with great fondness about his early days on the ranch. "Dad had an old high-wheeled Cadillac car that was about 90 percent used up. We would start driving from El Paso up the dirt road from Sierra Blanca to the ranch house and I could hardly wait to get there and saddle up those horses," Charles remembers. "We had a big old ranch house with a cooling wind porch all the way around it, and we'd sit out there and talk. I had a great time there."

Charles as a baby, held by his mother and anchored by his two sisters, 1911.

Early golf lessons from Aunt Lee in Cloudcroft, New Mexico, 1913. Charles never lost his love for the game.

They hunted quail and killed a big buck for meat every fall. In those days, because of low wages and uncertain pay (and maybe to seek out adventure or because of the old "greener-on-the-other-side-of-the-fence" syndrome), cowboys drifted around quite a bit. Some of them had their own small remuda of horses to work on. The Leavells had a big bunkhouse to shelter these itinerant hands during spring and fall gathering and branding times.

"There was a fellow named Kelly who was a hell of a good cowboy; in fact, he trained my favorite horse that Dad gave me when I was a little older—I called him Kelly Horse—and Mr. Kelly taught me how to ride, rope, and hunt with special skill. And every year I'd wait and worry until he returned.

Charles's father with his children and high-wheeled Cadillac, 1911.

The Leavell ranch house with Charles on the rumble seat of a Kelly horse being ridden by cowboy Kelly.

"I had friends my own age out there, and two or three of us would go out on horseback and ride down into the canyons that were always full of quail—blue quail. Those things could run like antelope. We'd get behind a whole covey and stay just far enough back so they wouldn't flush. They'd start running and finally when they were out of breath, they'd fly. And boy, we'd start spurring our horses chasing them down. They were flying and we were flying too. Over the rocks and through the cacti and the minute they lit we'd slow our horses and start 'em running again. Before you'd know it their wings were barely flapping and I'd jump off my horse and go grab a few quail and put them in my shirt. We'd take them home, wring their necks, pluck 'em, cook 'em and eat 'em. Boy, nothing ever tasted any better. Quail and rabbits and boiled potatoes were about all we had. I loved that place.

"We had a big old chuckwagon and a black cook named Sherman Grundy who would come during roundup times. Every year he'd bring four new wild mules to pull that wagon, and they usually ran away with it the first day of work, but he was a hell of a hand and soon had them properly trained.

"One time some men came up from where they had been hiding out in Mexico and rustled a bunch of Leavell cattle. Dad didn't tell the law, but he did tell a tough itinerant cowboy working for us at the time, named 'Babe,' of all things. He was a tough sucker who had a crippled arm. Dad trusted and liked him, and he told me to watch out for this fellow because Babe didn't take any foolishness off anybody. Back in the hills of Tennessee where Babe was raised they had a custom. If two young guys were courting the same girl, they'd tie the two together by one arm each and put a knife in their free hand. Then they'd fight until one was dead or crippled or both done in. That's how Babe got his disabled arm.

"Dad and Babe got a string of horses together with pack mules, rode down to Sierra Blanca and got deputized, tracked the rustlers across the Rio Grande border and caught them. They recovered the cattle and brought the thieves back to jail where they were convicted and sent to the pen." Charles says with undeniable conviction, "My father did that, and that always impressed me about the old boy."

Clarence Graham (left); Charles's father (center); Babe, the young man from Tennessee with the badly injured left arm (right).

Charles, Jr. was just as tough, at times, as his dad. One time Charles's father sent him and a young friend, by themselves, out to the ranch to paint the roof saying, "We must preserve those shingles." They had a fifty-five-gallon barrel of red paint, and were slowly getting the job done. It was unusually hot that summer.

Charles had been designated as the cook and this gave him some extra confidence. The boys would eat supper about 8 o'clock when it was a little cooler. Charles was wearily, but proudly, cooking the meal. He fixed biscuits and gravy and fried up some bacon and eggs. His friend, the son of a prominent unnamed man, said, "You know, Charles, I sure don't like your cooking."

Charles was extremely incensed and said, "Well, damn you to hell then if you don't like it." They argued a bit and Charles told the kid he could just do his own cooking from now on. The dispute grew to the point where the boys got into a fistfight. Charles hit the other kid in the face with the plate of the food that he'd found so distasteful, adding, "Swallow that, you ungrateful punk." Then things got serious; they really began trying to knock each other's head off.

One of the deputized cowboys who went to Mexico with Charles's father to capture the cattle thieves.

Charles recalls, "I remember he hit me so hard I went through the screen door and landed on my back out on the porch. He came and jumped on me and I knew I'd had it if I didn't gather up a lot of strength from somewhere."

He did. Charles threw his adversary over on his back, never thinking of his own bad leg, and started punching the boy's face into bloody welts. Then he started pounding the kid's head up and down on the porch and the boy began to cry.

Charles says, "I knew I had him then. I let him up without a word and he was sniffling and rubbing his face. He walked off down the dirt road toward the distant highway. There was a little plume of dust that rose where he walked. He finally moved out of sight over a hill. That's the last I saw of him and it was all over my terrible cooking."

Charles probably didn't know it at the time, but recovering from the powerful blow that had sent him sprawling surely helped instill the strong will that was already forming in him—the will that later allowed him to go out and conquer many parts of the world. He had been forged in the fire of the ranch cookstove. The hardness of steel in his belly always showed thereafter. A lot of people can attest to that with both rancor and admiration.

An *El Paso Times* article from the summer of 1926 reveals a lot about Charles Leavell and the relationship of respect he and his father shared. The headline reads "El Paso Boy's Nose Is Cut Off In Halloween Night Gang Fight." The article describes an attack on sixteen-year-old Charles Leavell, Jr., by a gang. In the attack Charles's nose was almost severed from his face. His father is quoted in the article, "I do not object to my boy fighting, and if he gets whipped in a fair fight that's his hard luck. But I do object to gang attacks, especially where it is evident that an instrument is used." Today Charles explains simply, "Those were wild times." He had countless, but different, wild times yet to come.

Charles had a grand time at the Figure Two Ranch until his father went broke during the Great Depression and died. However, Charles, Sr. left the family a $5,000 life-insurance policy and a big house in El Paso clear, so Charles's mother moved herself, Charles, and his three sisters into the house. There was plenty of space: seven bedrooms, three

A good string of fish caught by Charles (left) and his father in the Chama River in northern New Mexico, circa 1930s.

baths, and a big upstairs. But now something had to be done with the Figure Two Ranch, and it was up to twenty-two-year-old Charles to do it.

Nobody wanted the cattle, but he sold the ranch to the Hunter-Grisham Oil Company—one of the old drilling companies that later became part of Exxon—for $1 an acre. He took the $40,000 in cash to the bank and paid his father's debt, which was exactly $40,000. He says, "That ended that ranch. I've never been back. I never wanted to see it again. I loved it so."

He would work very hard for fifteen years before he would start looking for a ranch of his own to buy.

CHAPTER 4

Back When the Horses Came Running

There had been small, almost family-like, expeditions into the San Luis Valley (the 4UR realm) for a number of years before the first officially recorded one by Don Diego de Vargas in July of 1694. The pueblos had revolted in 1680; hundreds were killed and the Spanish were pushed all the way across the Rio Grande into Mexico. Twelve years later, de Vargas led an army north from the river to begin to reclaim the kingdom of New Mexico for Spain. By 1693 he had set up his army and government headquarters at Santa Fe.

The Utes had been trading south at Taos, Ojo Caliente and other points ever since they had adapted to the horse culture. When trade failed to work, they handily developed raiding parties to secure mounts and pack animals.

In July 1694, on the very day of de Vargas's expedition, Ute (called Yutas at that time) scouts had reported that a group of enemy tribes from the south was moving into the Utes' buffalo and antelope hunting territory. The Utes got along better with the Spaniards than they did with the Taos, Tano, Tewa, Picuris, Jemez and Keres with whom they were now at war—as they were periodically even before the pueblo rebellion. The Utes deeply resented the killing of the friars and made constant war on the pueblos, vowing to do so until they were destroyed for causing the Utes' friends to leave the kingdom.

Many of the pueblos' hunters had come into the Utes' buffalo-grazing lands. A few had been disguised with suits of armor and leather hats and jackets shaped to look like the those of the Spanish. Early on, this ruse had caused great loss of buffalo and Ute lives. This trickery is what the scouts thought, and reported, was so boldly about to occur again when they spotted the de Vargas expedition approaching. The Utes were, as incongruous as it might seem, feeling heated blood in their veins—the blood of battle and vengeance.

Some days earlier, on June 30, governor Don Diego de Vargas had set out from the village of Santa Fe where the one thousand inhabitants were nearing starvation. He was desperate to barter for foodstuffs and other supplies. He had about one hundred mounted presidio soldiers, thirty or forty Pecos Indians, a citizen militia and an unknown number of town officials for his little army. It is said that they had a remuda of perhaps two hundred extra horses for use and barter, as well as around eighty pack animals and their attendants.

They arrived at the Taos pueblo expecting to trade horses for grain, but the pueblo was eerily silent, with no human movement in sight. No doubt de Vargas expected an ambush, but after careful searches the party realized the entire pueblo had fled to a hidden sanctuary canyon in the nearby Sangre de Cristo (Blood of Christ) Mountains, knowing it would be foolish to take on such a large, well-armed force again. The de Vargas expedition spent four peaceful and profitable days at the pueblo shucking out and helping themselves to corn from the Taos granaries.

Indian scouts from Pecos had been reading fire and smoke signals around the Taos valley and had learned that several pueblos intended to ambush them in the mountains and canyons on the route back to Santa Fe. De Vargas probably thought he had the forces for victory, but knew that a fight would be too costly, both in men and especially in the corn they so desperately needed in Santa Fe. We must give de Vargas credit for listening to some of the leaders of his group and deciding on the route of trickery to the north before turning back south after crossing the Rio Grande to the west. This decision made history in and around the land of the 4UR, and part of that decision rested on the fact that they would be in the territory of their friends the Utes. So, wisely escaping one planned ambush, de Vargas headed straight for another mistaken one.

On July 9, 1694, de Vargas and his men found a wide, shallow crossing of the Rio Grande, about five miles south of where Colorado Highway 142 now crosses the Rio Grande between Manassa and San Luis, Colorado. On that long-ago day, the ford was pointed out to de Vargas by San Juan Pueblo Indians. The Tewas and the Tiwas had for centuries come here to mine turquoise and to hunt, for sport and food, the waterfowl whose feathers were crucial to many of their sacred ceremonies. (Three hundred years later—to the day—after long and careful research, a noted historian of the area, Ruth Marie Colville, who was ninety at the time, led a group of twenty-five or so to stand in de Vargas's footsteps and honorably and truly commemorate the momentous occasion on the vast land that surrounds and contributes to the 4UR saga.)

Moving on four leagues southwest to the San Antonio River, on Sunday, July 11, de Vargas set up a night camp. The Utes were watching from across the river and carefully planning their dawn attack on the supposed pueblo Indians. Governor de Vargas wrote in his journal of this attack:

> They charged with much force judging we were afraid and when we did not fire they still were convinced we were the rebellious Indians, their enemies. And thus our consideration put us in obvious danger, giving them more courage to come in among the shelters . . . of our royal army and troops. When all [the Spaniards] were armed and many mounted, some volleys of shots were finally fired at them because six of our own men had fallen wounded . . . and so with moderate discharge eight of them were killed. Realizing the strength with which resistance had been made, they decided to clear our camp, taking to flight, during which many were wounded because of the obstacle of crossing the river to which they had taken.

By then the sun was up, revealing the truth to the Utes: they had attacked friends. They waved a buckskin of peace, yelling "Anche," and "Pacuhe," which meant *brother* or *friend*. They rode back across the river and peace tokens were made. No regret was shown by the Utes about the damage done to them in their mistaken but understandable attack. Governor de Vargas wrote that he received and treated them kindly and gave them trinkets for hats, wool ponchos, knives, and ribbons. The rest of the men contributed gifts of corn and some dried meat. As was customary, de Vargas also gave a horse to the Utes' *Captain de la Guerra* in recompense for a brother lost leading the charge into the camp. The Utes helped to heal their own casualties and the wounded Spaniards with medicine made from herbs that grew along the banks of the rivers of the San Luis Valley.

Because their unusual herd of high-altitude buffalo numbered only about five or six hundred, it is all the more comprehensible why the Utes were willing to fight and die to maintain sufficient numbers for breeding; it was for maintenance of a large part of their tribe's lifeblood. After the battle, it made sense for the Ute wounded and their caretakers to head north to the most accessible medicinal hot springs, to heal their torn flesh. The rarefied air, water and game-laden mountains of the valley later to be known as the 4UR Guest Ranch were perfect for their needs.

The first colonizer of the New Mexico Territory was a wealthy Mexican, Juan de Onate, born in Guadalajara, Mexico, the son of a Spanish settler. He arrived with four hundred colonists and many horses, goats, sheep and oxen in 1598. Onate also sent exploring parties across southeastern Colorado into Kansas and as far west as the Gulf of California. Because of these explorations, and the royal confidence of the Spaniards, New Mexico Territory was thought to be much bigger than it actually was. Nevertheless, it included the San Luis Valley and the San Juan and Sangre de Cristo mountains

The San Luis Valley and the great mountains horseshoeing to the north were also the site of another historic event that changed this vast area of the West forever. That was Spanish colonial governor Juan Bautista de Anza's journey across the 4UR realm to confront the great Comanche warrior, Chief Greenhorn (more commonly known by the Spanish name, Cuerno Verde; in Comanche he was known as "Dangerous Man"). This head-on meeting, in 1779, settled once and for all the Comanches' power and position in the high plains and mountains where they had been driven.

26

Though noted researcher Ron Kessler and historian Phil Carson both agree that there will never be an exact reference date for this expedition, they have proved a very close approximation. It is known that de Anza gathered a force of some six hundred men, made up of professional soldiers, anxious settlers and allied pueblo Indians; it is believed that they were later joined by about two hundred Utes. The Comanches' daring and often brutal raids of both plunder and survival had made common enemies of all.

The expedition's power and resolve gathered near the confluence of the Rio Chama and the Rio Grande, north of present-day Santa Fe and Espanola, at the San Juan pueblo site. The cumbersome metal armor of the conquistadors had long ago gone out of fashion; it was too clumsy for the lightning raids of guerrilla warfare perfected by the Apache, the Comanches and later, to a degree, the pueblo dwellers. De Anza's pilgrimage moved through mightily varied terrain: across endless plains; up mountains over 14,000 feet high, with thin, clean air; down rugged, often seemingly bottomless canyons; across rivers and high mountain passes.

The governor had wisely chosen a much longer, and more difficult, route along the eastern flank of the San Juans to the west, instead of up alongside the easier eastern Sangre de Cristos where the Comanches had served up many ambushes. Comanche bands would enter Taos or Ojo Caliente to trade, as they had many times before, but then, as historian Phil Carson wrote:

> Treachery on one side or the other would provoke violence. The origins of hostilities were forgotten in the ensuing pattern of attack and counterattack. Sporadic records from the mid-1700s reflect occasional punitive Spanish forays against the Comanche across the plains of northern New Mexico and Colorado, but neither side won a decisive victory.

De Anza was determined to conquer the Comanches once and for all, so the territory could become peaceful and prosperous under his reign.

Finally, the firmly committed army moved out north past Ojo Caliente, Antonito, and what is now Alamosa, Monte Vista and South

Fork, just a few miles east from the paradise of the 4UR hot springs; on up through the Saguache area and the mighty mountains of Salida and Canon City; through and over Ute Pass, just north of Pikes Peak, and on eastward by the area of present-day Colorado Springs. Then de Anza turned back south toward the pueblo and surprised an encampment of Comanches. They attacked with much vigor and won the battle in short order, but soon discovered that these were not Cuerno Verde's warriors. Feeling both the exaltation of victorious battle and the disappointment of missing the main quarry, de Anza rallied his force's depression back into determination. By then they had learned that Cuerno Verde had moved south for another raid on Taos.

De Anza had, early in his career, distinguished himself in battle, earning a captaincy at the age of twenty-four. He was rapidly to assume command of the Sonora northern frontier at the presidio south of Tucson, at what is now a rich person's playground and artists' colony called Tubac. He had also valiantly made a trail from the San Gabriel Mission in California (near present Los Angeles) all the way from Sonora. He topped that adventure by moving two hundred and forty settlers and families north to Monterey, California, paving the way for the settlement of San Francisco Bay. Without doubt, both de Anza and Cuerno Verde were great warriors. There could be no impasse here.

De Anza's small army moved swiftly as possible south along what later became known as Taos Trail. They followed that track, paralleling Fountain Creek where it plunged into and enlarged the Arkansas River at the site of present-day Pueblo, Colorado. Then they turned southwest toward the often-used Sangre de Cristo Pass. However, to their great surprise, they met Cuerno Verde and his warriors returning from Taos long before any of them reached the pass, near the eastern edge of the mountains. (Today those hills are called the Greenhorn Mountains, supposedly in honor of the great chief who met his final battle there.)

The Comanches' raid on the Taos fortifications had been unsuccessful. Cuerno Verde's mood must have lifted when he saw an enemy out in the open. Had he accurately weighed the possible results, though, Cuerno Verde surely would have fled to fight another hour. It was not to be. Phil Carson, quoting from researcher Ron Kessler's reading of de Anza's journal, wrote:

They drew almost within gunshot firing off their own muskets. In this way was recognized from his insignia and devices the famous chief Cuerno Verde, who, his spirit proud and superior to all his followers, left them and came ahead, his horse curvetting spiritedly. Accordingly I determined to have his life and his pride and arrogance precipitated to the end.

De Anza, with his best soldiers, charged hard on horseback, successfully cutting off and trapping the chief in an arroyo. Cuerno Verde's horse was shot from under him but he leaped safely free. Protected somewhat by his horse's body, he fired both guns and arrows, in a final defiant death stand, at the same forces that had for so long tried to eradicate his people from the earth. De Anza wrote with much admiration that it was "a defense as brave as it was glorious." It was, of course, in vain, because the air was now thick with Spanish lead. The mighty chief fell, along with his followers.

The quiet that followed the brief but bloody scavenging of the chief's signature war-bonnet, as a proof of victory, was a sighing symbol of the last gasp of the great warring Comanches. In a short time de Anza and the Comanches would agree to a lasting peace, so that trade could begin for the area with the Cheyenne and the Arapaho.

This pursuit across the 4UR realm and the final battle are only a small part of the multitudes of adventurous and noble spirits, followed so doggedly by Charles Leavell, embedded forever in the great sphere surrounding the final destination of peace and beauty in the 4UR valley.

CHAPTER 5

The Launching of a Giant

Before Charles Leavell, Sr.'s death and the loss of the Figure Two Ranch, Charles, Jr. faced that period in life that begins to transform childhood experiences and training into either a life of brave dedication or one of timid excuses. Charles's father had been a tough disciplinarian and Charles was always expected to pull his own weight in the chore department while he was growing up. Besides his strenuous, but fulfilling, activities on the ranch in his boyhood, he became a strong swimmer and developed a powerful upper body that would be a great asset in his upcoming life of finance and outdoor adventure. The fact that he was a proficient swimmer, golfer, fisherman and hunter led to countless business connections that would have been impossible without this hard-won strength.

A couple of years before graduating from El Paso High School, Charles sought out part-time employment as an electrician's helper, installing turbines and such crucial things as transformer stations for Stone and Webster, a Boston company with extensive operations in El Paso. After he graduated second in his class at El Paso High, his father informed him that he had been accepted for entry into Dartmouth, Yale, Cal Tech and Stanford. He chose the latter and never regretted it. His interest in electrical engineering grew and that is what he majored in at Stanford. During the summer vacations, all through college until graduation, he continued to work for Stone and Webster of El Paso.

Charles says, "I studied in the fairly early stages of electrical engineering, designing transmission systems and all that." Then he adds

Charles Leavell.

with a sly, barely discernible smile, "Of course, I didn't go quite as far as a couple of my classmates, Hewlett and Packard."

Charles studied hard and worked even harder at his summer job, combining book and academic learning with on-the-ground experience from the very beginning. This gave him a tremendous advantage over most of his competitors as he moved on to his professional future. At Stanford he also found time to compete at a college level in swimming, and in spite of his physical handicap he made the second team, specializing in the fifty and one hundred meters. He was a strong member of the second team and also played water polo quite well. With his typical lust for all of life, Charles said simply, "It was a hell of a lot of fun."

By this time America was right smack in the middle of the greatest depression it had ever known. The senior Charles Leavell's illness had weakened him severely a couple of years before his death, so it was up to Charles to continue working even harder. He had paid his way through college and he was now the sole support of his entire family, consisting of his mother, his ill father, three sisters and himself—a lot of odds to overcome, but he did so without a single whine.

Upon his graduation from Stanford in 1933, young Charles decided he must make a decision about his direction in life. So after his father's death, and with enough money scrounged to take care of the family, he wandered around Europe for a year. He was fascinated with all the ancient, classic history and arts, but the long border between Mexico and the United States was singing that old desert song, and El Paso was leading the choir. He returned a little more polished, a little more worldly, but ready to get going right out of his own home town.

Charles says he didn't have any exact goal that he recalls, but thinks he might have already decided he wanted to be a contractor—a big-time contractor with big projects. And he made the right decision. He later stated, "I loved the adventure. I loved the risk. I loved the organizational challenge it took to bring in the exact right subcontractors. And I would begin to really bid against the 'big boys' of the country, and then I began taking them in as partners, and I expanded into three major branches in my business. One was general construction primarily—hospitals, hotels, you name it. Then I had a heavy construction division which was engineering on a large scale, lots of dams, bridges, nuclear facilities of all kinds." He adds, "But what was the most fun—and the most profitable for the time spent—was our ranching enterprises division."

Yes, this ranch child would conquer the building domain—a worldwide one at that—but what is remarkable is how often Charles Leavell uses the word *fun*. Only the wisest of paupers or kings realize that everything has been wasted and tossed into the great sandstorms of time if one lacks the will and wisdom to have true fun out of one's brief life on this particular planet. Charles has made an art form of fun, absorbing great joy from art and business, trout fishing and grand outdoor and indoor feasts, or just having a quiet visit with a close friend. He has done it while surrounded by the greatest of all American mountain ranges, the Rockies, which pushes holes in the sky from Alaska through Canada all the way across the 4UR realm to Central New Mexico.

Shirley Terrell, the woman who was to become Charles Leavell's wife—his love and his constancy of life—has, to say the least, an impressive family background. Her Houston-born father was a noted eye-ear-throat specialist who graduated from the University of Alabama and studied special treatments in Heidelberg, Germany. After his European studies, he moved to Dallas where he met and married Homer Callier. During World War I, he joined the British Medical Corps, later winning a Distinguished Service Cross. When the doctor returned to Dallas, they found his wife had tuberculosis, so they decided to move to El Paso where the dry desert air had long ago been proven curative. They moved into a house right across from the Leavell family home on Federal Street. This is where Shirley was born. Fate at work!

Prior to that marriage, Dr. Scurry L. Terrell was the first physician for President Theodore Roosevelt. Terrell was on the speaking platform with Roosevelt when he was running for president the second time—the time of the attempted assassination. Shirley recalls, "The remarkable thing about President Roosevelt surviving the shot was his very long speech, folded tightly in his vest pocket. The shot hit the speech and its velocity was diminished just enough to prevent a

Charles Leavell, age 10 (above), and Shirley Terrell (right), as a baby, in front of the Leavell house. These two photographs were taken at approximately the same time.

fatality." Shirley and Charles still have the life-saving speech with the bullet hole in it. Dr. Terrell followed through, going directly to the hospital with Roosevelt, and ended up going back to New York with him. He stayed on at the president's home for two more weeks, helping select specialists for his treatment and survival.

After Shirley was born, the Terrells moved to California for a while, where her mother died. Soon afterward, Dr. Terrell longed for the border just as Charles had and moved back to El Paso to take up his practice and marry again. He stayed there until he died.

There are a town and county in Texas named Scurry after some of Shirley's pioneer ancestors. Shirley's grandfather, Edwin R. Terrell, served in the Texas senate, and there is a Terrell County named after him.

Charles recalls this era. "Dr. Terrell was a very sophisticated man and highly respected by my family. Shirley had been more or less adopted by her mother's younger sister, Nix Callier." Aunt Nix lived in Beverly Hills on Roxbury Drive, with much taste and elegance surrounding her. Shirley went to El Rodeo Grammar School in this city of opulence and then to Radford School for Girls in El Paso. Then she went to the elite Hockaday Boarding School in Dallas and from there to study at Villa and Chateau Brillaumount, Lausanne, Switzerland, returning to graduate from Pine Manor College, Massachusetts, in 1937.

When she came back from Switzerland, the summer before going to Pine Manor, Shirley met Charles for the first time. She didn't pay a lot of attention to him because in her youth he seemed a bit old to her.

One Christmas season, Charles's mother did a little matchmaking. He was in his mid-twenties now and dating several girls, but none he wanted to lay down his life for. His mother said, "You've just got to meet this beautiful girl, Charles." He did. His mother was right. He was taken with her beauty and grace at once, even though he thought she might be a bit young for him.

Shirley had to go to Dallas for a coming-out reception and he vowed to pursue her when she returned in spite of her youth. He found out that there was lots of heavy competition for her dates and her hand. Several handsome lieutenants from Fort Bliss and some of the better male catches of El Paso were after her attentions.

Charles' everlasting adoration for his lovely wife is evident in his voice as he speaks of her. "She was extremely popular and I took her out this night just knocked over by how cute she was and how dear. She had such big blue eyes, a beautiful figure and a wonderful sense of life and adventure. We danced a little and I said to myself, 'Boy, that gal has got it. She is something.' But I had to peel off those other dadgum suitors. I had to get rid of them. There were two men from California and one lieutenant here that were tough competition. It was an exciting courtship. So I took her out all determined to finish it off properly. I took her to this barbecue place out in east El Paso called Tony's Beer Joint. I didn't have enough money to order fancy Scotch whiskey or whatever drinks she was accustomed to, so we had beer."

Shirley continued painting the scene of their engagement. "That Sunday afternoon, as I recall, Charles said, 'Will you marry me—'—this was in the fall—'Will you marry me next March—' "

Charles added, "She said 'yes,' then wanted to know if I was sure we were going to be all right financially and everything. And naturally I said, 'of course.' " Charles's eighty-four-year-old blue eyes sparkled like liquid diamonds as he remembered this.

Shirley Terrell went on to Dallas and made her debut, adding, "So when I came back from Dallas why, we just put it all together and got married that year of 1938."

Charles takes his turn, "We had a wonderful wedding in their family home near ours."

Shirley wraps it up very simply with, "And we've been together for fifty-eight years now."

They look at one another almost coyly, as they did in their youth, with the deep affection and silence that only a long life of wondrous struggles, hardships and glory shared together can bring about.

That year, 1938, was the time of beginning—their time, the kind of time that has nothing to do with clocks or calendars.

Charles took a job with Peterson Lumber Company and was soon promoted to head of the roofing department, with a bonus arrangement. Every time he took in $100 on a contract he received $3.00 extra. So, he was working as hard as he could, adding as many bonus threes to his salary each month as possible.

Charles reluctantly tells this story, "Anyway, I had this little desk, a phone, a pad and a pencil. The phone rang and this lady said please come over immediately and roof my house. I'll never forget 2429 Sacramento Street—not now and not ever in the future.

"It was early in the morning and I went tearing out. I gathered my two wetback helpers and we fired up my little black tar-pot and put it on the back of my pickup ripping off to roof this house. Ambition was whipping me across the back like a muleteer.

"We did a beautiful, fast and efficient job. The next morning I was at my little desk, oh ever so proudly writing up the order, the bill and everything when the phone rang. The same lady said, 'Why didn't you come roof my house—' I said, 'I did. Go out and look.' And I thought, What's the matter with her anyway—

"After checking, I found that I had become so excited about the order that I took down the wrong house number. I had roofed at 2409 Savannah Street. The people who got the new roof had been away for a day or so and just accepted the whole thing. I got my ass fired out of that place, nevertheless. I made a lot of mistakes in my time, some of major proportions, but none stayed with me like that one."

Charles looked around for some building jobs or electrical contracting that he could do on his own, and found some, but was short on the funds needed to go into business for himself. So, *they* went into business.

Shirley had an inheritance of her own and she loaned him $3,000. He recalls, with that wondrous, bemused smile, "She made me sign a note and I had to pay her interest."

When Charles told his good friend, Lee Orndorf, who was an expert at handling money, about this family transaction, Lee told him, "You gotta learn how to borrow money from a *real* banker. Now don't be afraid of debt. Learn to use it."

Charles followed his instructions and that was another thing he never forgot. Mr. Young was the president of El Paso National Bank. Charles told Mr. Young that he needed $1,000 and that Lee Orndorf would co-sign the note. He paid the interest on the note and soon paid back all the money. He did this five different times. The sixth time, just as his friend had instructed, he borrowed without the extra signature.

That's how he established his first full line of credit that in later years would be good for $600 million—in today's terms, about $2.5 billion.

He went to work and built a porch and a garage on a fellow's house, and then did five "spec" (speculation, not presold) houses on the money he had borrowed from Young. But then the war came on and there was a sudden freeze on building materials. Before Charles could possibly scrounge together enough scarce material to finish his houses, Mr. Young called in his note. Charles talked to him with much earnestness, but the best he could get was, "I'll continue if you'll get your father's old friend, the mayor, to sign your note."

Charles told Young he could never do that. He was not a beggar. And Young told him, "Well, get the signature or suffer accordingly." The houses were half-built. Charles was desperate. He actually sat down on the curb and cried a moment. Then, as always, Charles gathered his natural courage, marched over to the State National Bank and demanded to see a loan officer named George Matkin, who would later become chairman of the board. Mr. Matkin asked, politely, considerately—as most truly great gentleman seem to do—how much Charles needed. Charles told him and sweated blood and part of his soul as Mr. Matkin excused himself and departed the office. After many agonizing eternities Matkin reentered the room, bringing with him a message that would move Charles H. Leavell on through multitudinous successful lives of finance and joyful adventure, without which he could not have maintained his dream of the 4UR.

"I'll tell you what we'll do, Mr. Leavell, we'll let you have $30,000 to pay off that other bank and another $30,000 to finish off those houses."

Charles banks with State National Bank to this day. He served on their board and Mr. Matkin served on the Leavell Company board. When Charles's company grew into a truly huge contracting firm, they sent Matkin to New York to help arrange the massive financing it would take for Charles's worldwide endeavors. Oh, yes, and he paid off the loan to his wife—with interest.

He had tried very hard to get into some form of military service, but his leg prevented that. It was the military's great loss and Charles's eventual great gain—although his military contributions were about to begin in a big way. There were many "hurry up" war contracts and military commitments that had to be built with all expediency. Charles bid on an Air Force base construction at Deming, New Mexico, and won. As he was packing his bag, Shirley asked when he would be back. That was a Wednesday and he said, "Monday." He was gone for a year and several other construction jobs before he made it back home.

He had joined in with Lee Moor and the Mayfield brothers, Tom and Davis. At the opening of bids at Corps of Engineers' Albuquerque district office, the colonel read the competitor's bid first. It was a million dollars higher than Charles's. In a panic he tried to retrieve his bid before it was read, but he never made it. When the job was finished, Charles had netted $500,000 for himself and made everybody concerned happy.

Shirley knew and understood the importance of the project to the Leavell family and to the country, or it would have been very difficult for her to accept this kind of absence during the early years of their marriage. Many decades of exploits later, Charles was quoted thusly about his wife: "Shirley has been my guiding light ever since I married her. She has been my confidante, advisor, critic and pal. We've had many bumps, but darn few have we created one on the other. It has been a magnificent relationship. I depend more and more on her judgment. She is a great observer of people, their integrity and depth of interest in the company. It was largely through her keen perception of people that enabled me to establish profitable and enduring relationships. And she has enjoyed every minute of it."

Shirley later served on the board of directors of the Leavell Company and often relieved the tedium of business talk by asking Charles, in the middle of a meeting, "When is the board going to vote the stockholders a dividend—" It always caught Charles off guard and pulled a round of laughter from all. Yes, the Leavells knew how to work, but they have never, not to this day, forgotten the fun that makes it all worthwhile.

Their two children, Mary Lee and Pete, also adapted and applied this same philosophy to a great degree. Since the time when the four of them took their first vacation together at the 4UR, with all those bedrolls, camping and fishing equipment and gasoline strapped on the roof of the car, they accomplished far more than most families are ever privileged to achieve—but the Charles H. Leavells were just tuning up for the fascinating future they would create. Without those two years of sacrifice, separation and success, they would never have made that special trip to Colorado, and this great dream would never have been related.

CHAPTER 6

Consummate Courage,
White Death and Making History

Roughly, and specifically at the same time, the 4UR realm's four corners are pegged on the northwest corner by the Black Canyon of the Gunnison National Monument near Montrose, Colorado; on the southwest corner by Mesa Verde National Park; on down to the Aztec Ruins National Monument near Aztec and Farmington, New Mexico, looping back to include the Chama, New Mexico, valley up to Taos as the southeast corner; then north again to Questa and Costilla, New Mexico, to San Luis, Fort Garland, Gardner and Westcliffe, and on to Pikes Peak near Colorado Springs at the northeast corner; then back west to Montrose. This is truly a land of unsurpassed adventures, endless legends of courage and suffering, horror and heroics. Even the ill-fated expeditions were of enormous benefit to those who followed.

Why has the description of the 4UR realm—the San Juan/Sangre de Cristo/Chama/San Luis Valley domain—been enlarged from the first general description? Because Taos played such a major role in the entire region for hundreds and hundreds of years. True, the fingers of the great valley are cut off from Taos by the mighty Rio Grande gorge. Nevertheless, Taos was of such importance to the outfitting and launching of initial exploration that both history and understanding would be incomplete without it.

The Mesa Verde ruins near Durango have become one of Colorado's top tourist attractions, and present a never-ending struggle

Wolf Creek Pass, south of the 4UR Ranch. (Courtesy, Adams State College Library)

Near Creede, northwest of the 4UR Ranch. (Courtesy, Adams State College Library)

and object of study to archaeologists trying to understand what happened to the inhabitants of these majestic cliff buildings. The other, more nomadic Indian peoples of the area called them *Anasazi*, which to them meant "enemy ancestors" but to later pueblo tribes and others meant "the ancient ones." No matter. The wonderful stone houses in the cliffs were abandoned somewhere around 1300 A.D. because of a great drought and subsequent migration, we are told by many scholars. But the Anasazi left beauty in the surroundings, and their spirit still permeates this wondrous, mysterious place, causing each visitor to daydream and wonder, just a little, about the destination of the departed.

There was the Alferd G. Packer expedition in the winter of 1873-1874, which became entrapped in the one of the San Juan range's mighty storms and whose leader was accused of eating five members of his party to save himself. (There were fairly solid claims of finding knife marks on the bones of those he butchered.) At any rate, the stories have been endless, and countless poems and musicals record his "dastardly" deeds. When Packer was being tried at Lake City—a town near Creede—the judge asked how many Democrats were on the jury. The jury foreman answered, "Your Honor, there ain't none. Packer done et 'em all."

The great chunk of granite that pegs the northeast corner of the "realm" is the world-famous Pikes Peak. It was named for Zebulon Montgomery Pike, an American soldier and explorer who was born at Lamberton, New Jersey, in 1779. He received his education in Easton, Pennsylvania. In 1805 he entered upon an expedition to locate the source of the Mississippi River; next he spent time exploring the territory of Louisiana, to pin down the southern boundary of the Louisiana Purchase; and then he moved on to Colorado and found this massive mountain spearing into the sky. When he first spotted the form of the peak from a great distance, he called it "a small blue cloud." Days later, he tried to climb this mountain and failed. Nevertheless, because of the way it stood out from the vanguard of the Colorado Rockies, and the manner in which Pike described it in his journals, it was named after him.

At about this time, in 1806, several Spanish expeditions were sent out across the plains country of the San Luis Valley and the nearby environs. These forces were charged with a double purpose: to improve the erratic relationship with the so-called Plains Indians and to see what the Americans were doing.

For many years the Spanish rulers of Santa Fe had heard reports of the lucrative adventures of trappers and hunters in the lower San Juans. Later beaver trapping slowly emerged as a strike, not unlike the later ones of silver and gold, because of the foreign and eastern demand for heavier coats and hats. The Spanish wanted to know how these mountain men were acquiring these reported riches. No full reports of what they found are readily available, mostly because these largely undocumented expeditions had no well-known, "name" leaders.

That year these same Spanish rulers of Santa Fe also heard about a group of soldiers about to enter the inexact northern border of the

New Mexico Territory, led by Lt. Zebulon Pike, U.S. Army. Everywhere the U.S. soldiers traveled they spread the rumor that they were there principally to collect botanical specimens, roughly map the landscape and (just like the Spaniards) establish a friendship with the Plains Indians. They also avowed that they wished to establish trade with both the Spanish and the Indians, though many believe that this was merely a cover for a purely military expedition. After finding the peak that would forever bear his name, in November 1806, Pike's expedition wandered around the San Juans, through December and January into 1807, suffering mightily from the cold of this great range to which they were totally maladjusted. They became thoroughly lost and were in great danger of perishing from avalanches or simply starving and freezing to death. However, (unlike Fremont's fourth disastrous expedition), the angels of chance allowed them down into San Luis Valley, which was having a rare mild spell.

Pike pitched camp in what he supposedly thought was American territory and raised the American flag. In February 1807, he gave instructions to the company doctor, John Robinson, to go to Santa Fe alone. The doctor carried a bill of indictment with him for a previous messenger—a French trader named La Lande who had failed to return. Other than that, Dr. Robinson was supposedly sent to find out how far the camp was from Santa Fe. Lt. Pike's company waited anxiously, not knowing if the doctor had fallen victim to a sudden storm, the many Indians of the area, or the Spanish themselves.

It was about three weeks later that a company of heavily armed but polite Spanish reached the Pike camp and pointed out that they were camped on the Conejos River just above the junction of the Rio Grande. Ironically, this was very near, if not on, the exact spot where de Anza had camped when attacked by the Utes a few years earlier. The Spaniards gave the Americans extra provisions in a gesture of courtesy, while at the same time politely ordering them to accompany the Spanish to Santa Fe as prisoners. They were informed that Dr. Robinson had already been sent to Chihuahua in northern Mexico to stand trial as a spy and that they would soon follow. After a stopover in Santa Fe, where they were wined and dined like world leaders, Pike's group completed the trek to Chihuahua in April. They were still treated well, although visions of eventual imprisonment or execution must have passed through their minds now and then. To their relief and surprise, the entire company was returned north into the territory of the United States unharmed.

There has been a difference of opinion—as with almost all things historical—as to Pike's true purpose. His journals reveal an astonishing amount of information, from the price of coffee and lead to the number of authorities, soldiers, and other items, in such great and accurate detail that we'll probably never know if he was truly there for the stated purposes or as a spy with the idea of later conquest. Pike died in 1813, during the War of 1812, but his name on a mountain he couldn't climb and his (purposeful or accidental) trip into New Spain gave him earthly immortality, and he has become an indelible part of the history of the great region surrounding the 4UR.

Another adventuresome explorer was John Charles Fremont. It is amazing that he was ever heard of in written history. (It would be especially so if he had been one of those who blame all their failures on childhood abuse by their parents.)

He was born on January 21, 1813, in Savannah, Georgia. His father was a French immigrant named Fremon. His mother, Ann Beverly Whiting, created a huge scandal by leaving her elderly husband, John Pryor, to run away with her lover, John Charles's father. Pryor petitioned the Virginia legislature for a divorce, but it was never recorded, so Fremont grew up as a bastard in a time when the word itself was a sacrilege and the fact a disgrace in a family. After his father, Charles Fremon, died in 1818, the "t" was somehow added to his last name.

His mother, with a family that by then included a second son and a daughter, moved to Charleston, South Carolina. As their finances slightly improved through "business" efforts of his mother and outside work of his own, John Charles was prepared for the college of Charleston. In 1829 he enrolled as a junior in the scientific department. Although a bright and attentive student at times, he suffered another destructive indignity by being expelled three months short of graduation for "incorrigible negligence"—whatever that was.

It is necessary to give a brief history of this unusual man to fully understand what happened to him on his fourth expedition to the West. That venture strangely touched the footsteps of all who trod upon the ground of the San Luis/Sangre de Cristo/San Juan range, and actually passed over the land that Charles Leavell would later own as his second-favorite ranch.

Because of the skill Fremont had shown in mathematics, a political leader secured for him a position to teach the subject on a U.S. Navy war sloop. The main focus of his astounding and erratic career really began when he left the Navy to become a second lieutenant in what would later be called the U.S. Army Corps of Engineers. In 1838-1839 he was accepted into John Nicollet's expedition to the plains between the upper Mississippi and Missouri Rivers. The two years he spent out there was his Virginia Military Institute, Stanford and Harvard, as far as education goes. He learned through practical experience sophisticated methods of surveying, including using the barometer to measure altitude, how to properly manage an expedition and how to make astronomical observations.

In 1841, Fremont made the most successful move of his life when he married Jessie Benton, the seventeen-year-old daughter of Senator Thomas Hart Benton. She was a writer and helped mightily in Fremont's five major expeditions to the Far West, particularly in the crucial reports on soil fertility, Indian villages and trading posts, as well as the almost uncountable adventures during his treks.

In the spring of 1840, John Charles Fremont was ready for his first major command and was ordered to survey the Platte River up to the head of the Sweetwater. He was accompanied by twenty-one accomplished men, mostly Creole and Canadian, but also including such figures as Charles Preuss, the skilled German cartographer, and Lucien B. Maxwell as a hunter. Maxwell, a man who once controlled the 1,714,764-acre Maxwell land grant near Cimarron, New Mexico (which reached into southern Colorado), made history of his own. The huge holdings were originally obtained as a Mexican land grant by his father-in-law, Charles Beaubien, in 1843. Even though Lucien Maxwell thought he owned, and did control, the massive grant, there was constant litigation over its validity until twelve years after Maxwell's death, when it was declared valid by the U.S. Supreme Court in 1887.

Fremont had as a guide another famous personage of western history, Kit Carson, who would become Fremont's close friend. Altogether, they made up quite a powerful group.

Fremont's first western expedition reached South Pass, another region of the 4UR realm, then headed northwest to—and scaled—what was later named Mount Woodrow Wilson, one of the highest peaks on the Wind River Mountain chain. By the time he returned to the St. Louis headquarters, fourteen months later, he had made an almost complete tour of the entire West, the Great Salt Lake and Fort Vancouver and had led a dangerously foolhardy midwinter crossing of the Sierra Nevadas in the vicinity of Carson Pass. They almost starved into a frozen death but miraculously survived. They healed up and resupplied at Sutter's Fort on the Sacramento, then pressed on to Oak Creek Pass, where they followed the old Spanish trail until Utah Lake necessitated a departure. Then the group moved on to the Wasatch and Uintah Mountains.

After his wife, with much foresight, compiled most of his reports, and Fremont rewrote them into readable form, the Fremonts (with the aid of Preuss) drew maps that changed forever the entire picture and concept of the West. This journey alone would have put Fremont in the permanent journals of western history. Several commercial editions of his report appeared domestically and foreign. John Charles Fremont, the little five-foot-two bastard, had become a national hero. He was made a captain, then a lieutenant colonel, as he moved ever upward and outward in work that caused improbability to become possible and then often reality. Fremont's wife wrote him, "They say that as *Robinson Crusoe* is the most interesting fiction of travel, so Fremont's report is the most romantically truthful."

On other expeditions, Fremont helped free upper California from Mexican rule—or took it away from them, if you like. Fremont had many stratagems. He was a major general by the time the Civil War broke out and in that capacity made a proclamation to free the slaves of owners disloyal to the cause. This made him a hero to Harriet Beecher Stowe and other anti-slavery workers, but the publicity, as it often does, backfired. President Lincoln removed Fremont from his command because he had acted without authority in a civil matter.

Many, many books have been written about Fremont, who (sometimes rightfully, sometimes wrongfully) had such a vast influence on the outcome and timing of the development of the West. One of these trips, the most important to this book, was his fourth famed and disastrous expedition of 1848-1849, which criss-crossed the 4UR realm. The driving force behind this ill-fated expedition was the need to explore, find and map a central railroad route to the Pacific. Noted San Luis Valley historian Ruth Marie Colville, of Del Norte, Colorado, said, "The men of Fremont's fourth expedition constituted a strange mixture, including mountain men, a German map maker, [a] French Canadian, a blacksmith, a Scottish captain of the Hussar, Philadelphia artists, a medical doctor, a freed black man, a tubercular, [a] mid-shipman, three Indians, a Mississippi gentleman and a teenage boy."

Fremont's expedition left St. Louis on October 3, 1848, with 35 well-equipped men and 120 horses and saddle and pack mules. At El Pueblo they shelled and sacked 130 bushels of corn—certainly, it seemed at the time, plenty to carry the animals through the Sangre de Cristo and San Juan Mountains or in an emergency. They were professionals and prepared.

The goal was never achieved because, as they entered the 4UR area in December, they ran into an unusually severe winter of terrible storms. However foolhardy Fremont had been in crossing the Sierra Nevada and nearly losing all his men, this was not the case here. The ordinary forces of unpredictable nature doomed this foray, as did the still-puzzling directions given by the famous, rascally, Taos mountain man, Bill Williams. Some blame this disastrous venture on Williams; others insist it was the fault of Fremont himself. Patricia Joy Richmond was no doubt correct in writing that Fremont would probably have preferred Kit Carson, his guide on previous successful expeditions, but Carson was occupied at the time between Washington, D.C., and other expeditions to the West.

As Fremont's group entered the eastern edge of what is now Wetmore, Colorado, a member of the party, George Hubbard, looked up at the forbidding winter Rockies and said, "Friends, I don't want my bones to bleach upon those mountains amidst that snow." His words aptly predicted what would soon be truth for many.

The snows kept coming, increasing in depth and consequently sapping the energy of men and animals, who were forced to camp in deep snow. There was no water. The corn that had been so confidently prepared for the San Juans was now being used up traversing the Wet Mountains. By November 29, the Wet Mountains were to the north and they could see the little Spanish Peaks to the south. The strenuous, roller-coaster terrain covered with deep snow exhausted the men as they approached the valley of the Huerfano River. They were at this time just south of the land of the J M Ranch—later to be owned and loved by Charles Leavell; in fact, they were looking right across it. They followed mountain man Antoine Robidoux's wagon road along the north side of the river and camped at the present-day settlement of Red Wing.

The next day the men and animals struggled through the canyon of the Huerfano for a few miles, then followed the road over pinon-covered hills. This place, where both the Pike and Fremont expeditions rode, would also be ridden by Charles Leavell, his family and close friends; part of the invisible connections in a web constantly in the making.

The 14,000-foot-high, snow-draped peaks of the Sangre de Cristos were hidden from their view. For this short time, the natural optimism of all great adventurers returned, because of the easier traveling. Little snow encumbered their climb, though piercing winds caused discomfort. They topped out on the 9,175-foot-high summit of Robidoux or Mosca Pass at noon on December 3, 1848. They were standing within the future boundaries of the J. M. Ranch and the 4UR realm.

Fremont and his men looked down into the valley of the Rio Del Norte. The Rio Grande and the plains surrounding the area (what is presently the hub town of Alamosa, Colorado)—everything looked a dreadful white to them, white plains surrounded by white mountains. Even the air itself would soon turn white. As it turned out, the old cliché "blinding" white proved horribly real. The Mosca Creek Trail—on Leavell's J M Ranch into the San Luis Valley—was extremely variable in its difficulty.

The next day they moved between the great sand hills and the foothills of queenly Mount Blanca. But the temperature was below zero, and deep snow and high winds almost stopped the group's movement altogether. Fremont, Preuss, the guide, Bill Williams, and a few other examined the gap at Medano Pass (still on Charles Leavell's J M Ranch). Preuss wondered why they had taken the route over Mosca Pass, which had used up more than a week, dangerously consumed supplies and exhausted the men and animals. The question will never be answered. Was Williams confused by snow blindness? Did Fremont's ego cause him to risk all? Had the easy access to the pinon hills somehow fooled the tough old pros? Or had fate simply taken control? At last Alexis Godey, who Fremont felt was a young Kit Carson, led the expedition directly over the sand hills through six feet of snow, following almost exactly the snow-buried trail made by the Zebulon Pike Lost Expedition four decades earlier.

Respected historian Patricia Joy Richmond wrote, "Five years later [after this disastrous fourth foray], according to Solomon Carvallo, the fifth expedition traveled up the San Luis valley crossing the Rio Grande del Norte, and entered the Saguache Valley." If this is so, they passed right next to the 4UR itself.

It had to be in Fremont's mind, stronger as each cell-freezing day weakened the expedition, that something was seriously wrong, but everyone knew that old Bill Williams was an experienced mountain guide. The problem that apparently no one—including Williams himself—recognized was that he had always entered this country from the south instead of the north; the different approach changed his view and remembrance of the landscape. If Williams had led Fremont up the often-recorded Canero or Saguache routes, they would by then have been trekking across the Cochetopa Park, heading for California as Bill had earlier so carefully explained and planned. Some historians suggest that Williams did recommend using Cochetopa Pass. Currently the mystery seems unsolvable.

So near, yet so far . . . another ancient refrain of mistakes. The Fremont party was trying to top out only a few hundred feet above their destination on Mesa Mountain, near the present town of Creede. They might as well have been in the center of Alaska. The mules and horses were reaching the point of last endurance, and the pint of corn rationed each day was practically useless. They wandered around, up and down, the upper parts of their bodies appearing and disappearing in the deep snow that hid gullies, boulders and fallen logs. The party must have looked as though they were walking on a snow-covered world vibrating with unending earthquakes.

Young Godey killed two elk that temporarily helped the humans but did nothing for the mules. The wind howled and pushed so savagely that it felt like foot-long frozen needles by the thousands were being driven through their bodies. To some that feeling would soon become an actuality—they would die trying to breathe air of ice.

By December 12, the expedition had struggled along the sheer rock-walled valley of Cave Creek into the La Garita Mountains from where they had left the Saguache. This was almost directly across the Rio Grande from today's 4UR ranch, in unimaginably rough terrain. It was an unholy struggle up and down mountains, climbing as much as two thousand feet in drifts sometimes fifteen feet deep, with humans actually breaking trail for the feeble mules. By sunset they were in a new camp at the west foot of Boot Mountain. Remarkably, all but eight mules and six men made it to that camp. The weather had moderated and for a time all exulted in the success of just staying alive.

How futile even to think about predicting Mother Nature's moods. By nine o'clock that night, fast-falling snow was blowing over everything: the grass, the men, the mules, the few surviving horses and their owner's souls. The eight missing mules had died in the tortuous attempt to cross Boot Mountain. Seven more, including Godey's personal mule, died before morning. The winds in that part of the 4UR realm can plunge the chill factor to seventy-five degrees below zero when they sail and twirl across frozen ice and snow. Deadly beyond mere description.

The remaining men started naming their camps, with a macabre sense-of-survival humor. At this time there was very little chance of them knowing where in holy hell they were, but on December 16, 1848, they were in a place they called Camp Dismal, somewhere in the vicinity of Wannamaker Creek.

The day was disastrous. Along their trail were scattered packs, saddles, blankets, pads and dying mules, which lay gasping their last desperate breaths in twenty-below-zero air—without the chill factor. Some, driven mad by the endless stress, plunged from the ridges into snowy depressions as deep as a hundred feet. Later explorers would find the bleached, lichen-encrusted bones along the ridged terraces of Groundhog Embargo and Rincon Creeks.

From above the waters of the Wannamaker, tough old Bill Williams lay down on one of the summits and wanted to die. They turned back in retreat to Camp Dismal. Richard Kern said that another half-hour on this particular furiously pummeled summit would have wiped out the entire expedition.

The next morning the storms abated enough for them to move out again, but each morning the men used tin plates to scrape several inches of new snow from their bedrolls. They were falling prey to desolation and despair of purest horror. As of December 20, only 59 of the 120 work animals survived, and a great many of those died in the next few days around the summits of the Wannamaker, scattered across the mountains like a giant buzzard's discards.

The men finally realized that eating mule meat was their only chance of survival. Many years later Thomas Breckinridge wrote that they had dined on baked mule, boiled mule, and many other delicacies concocted of mule meat. Of course, Fremont learned and remembered this lesson well for his fifth expedition, but they must have all been overcome with a snow-blind madness not to have thought of it sooner. But then, when the body is weakened by hunger, and is cold almost to the point of freezing solid, and is mostly blind from the blowing white, the brain might not work as well as that of a warm, safe person.

They had wandered from La Garita Camp, on the creek of that name, to Inscription Rock Camp, West Benino Creek Camp, Groundhog Camp, Preuss Camp, Christmas Camp, Holy Hell and on and on. On December 30th Ben Kern wrote that Camp Disappointment would be behind the expedition as soon as the bedding was moved. It was not all bleak, as luck and hunting skill sometimes combined to give them temporary feasts of buffalo steaks, meat pies, soup and coffee from their few remaining supplies.

Fremont had sent a rescue party to Taos with the best of their supplies and few surviving mules. This seemed to be his only choice, if for no other reason than to give the survivors something to look for and forward to.

By January 11, 1849, the men were beginning to stray apart. The dissolution of the expedition was now obvious. The men scattered out in two and threes, trying to head for the San Luis Valley to meet the overdue relief party from Taos. McGee described the fierce winds thus: "We would have to lie flat down, at times to keep from being swept off."

Two days earlier, on January 9th, Raphael Prone, a senior member of all Fremont's expeditions, lay in the snow dying from cold and extreme weariness. This tragedy of the death of his dear companion convinced Fremont that if he did not personally start for Taos the entire expedition would expire just as Prone had. Therefore, on January 11th, Fremont, with Alexis Godey, Charles Preuss, fourteen-year-old Theodore McNabb and Jackson Saunders, left for Taos with orders for the rest to follow and camp as best they could until rescue arrived.

Bill Williams had somehow gotten the original rescue party lost over and over. In one group, an Indian named Manuel begged to be shot because his feet were frozen solid. He was at first reported dead, but he survived and was later rescued by Godey. Another group below watched as Henry Wise lay down on the ice on a creek and died.

Fremont and his men finally arrived at some settlements above today's Questa, New Mexico, about forty miles north of Taos. Flour and goats were made available to them. Alexis Godey moved back north to the settlement of Rio Hondo to get healthy mules to transport supplies to the desperate men in the San Luis Valley. At last, relief was actually on the way.

When Godey arrived in the area, he found only one lodge and its contents, a trunk with some of Fremont's clothing and his instruments. It took Godey two days to find a man named Hubbard. He was dead but his body was still warm. Godey never let up until all the living were recovered. Two months after their climb into the San Juan Mountains, Fremont and his group and the sixteen other survivors were finally reunited in Taos. Most of the mules and horses had made fine food for predators. Though history calls this the Fremont expedition, from all written reports it appears that Lt. Godey saved the expedition's survivors, including Fremont, again and again.

The 4UR realm gives about eight months a year of green, plentifully watered paradise, but for a person without shelter—like Fremont—the other four months can turn to a frigid hell. Yet these very same storms have now turned Wolf Creek Pass, Purgatory, Telluride and other areas into a winter skiers' paradise. Zebulon Pike and John Charles Fremont may be sitting around a campfire in some other dimension, either ruing or laughing at this little twist of earthly fate.

It is just as amazing that historian Patricia Joy Richmond would sacrifice twenty years of her life to find the remains of camps and disaster areas over such a huge area of the 4UR realm. In her book, *Trail to Disaster*, she relates, "I traversed mountains, valleys, passes, rivers, and trails by four-wheel-drive vehicle, snowmobile, horseback, snowshoes, and cross-country skis; but most commonly by foot . . . while bearing a pack laden with cameras, lenses, film, books, maps, notepads and necessary survival gear." She checked and double-checked locations, maps and journals for all those years. Without her mighty on-the-ground efforts, the trail would have been forever lost to nature and guesswork, and the history of the 4UR realm would have a large void in it.

CHAPTER 7

The Days of Surging Power and Overcoming

By 1946 Charles Leavell was moving out and upward, but the long absences from his family were the hardest part—for all of them. Mary Lee and Pete were healthy, growing children of great spirit and he hated being away from them. The phone calls and constant letter writing helped, but it was Shirley's wisdom, understanding and faith through these absences that kept the family together and on even ground. Her constancy is a tribute to her that Charles has never forgotten and it has paid off in togetherness for decades.

Of course, Charles was absent by necessity, not by choice. The experiences of the Great Depression and the loss of the ranch he had loved so deeply during his boyhood imbued Charles with a conservative fiscal nature.

The early years of his chosen career saw bigger and better work. He hired Mary Jane Roberts right out of business college as his payroll clerk and general executive and traffic director. She was extremely cool under fire. He had well-chosen, highly qualified estimators and accountants. All in all, a tough team was in place. This was the crew that entered into the worldwide fray of contracting for the biggest roll of the dice outside of actual war.

In the years from 1946 to 1960, the Leavell Company took on forty seven projects, all of which were completed and built during that period, including Los Alamos and the White Sands Missile Range. This beginning work totaled tens of millions.

White Sands is, of course, where the first atomic bomb was exploded in 1945, changing the world forever. Now it is one of those places beyond proper description for those who haven't actually been there. Over uncounted millennia the surrounding chain of gypsum mountains eroded into a vast base of gorgeous white sand hills. Past all the military installations and the National Park Service gate is a glowing world of white sand unlike any other; it engenders a feeling of awe that borders on the divine. The people who come to this wondrous natural phenomenon from all over the world feel that their souls have somehow been cleansed. Many major motion pictures and commercials are shot in this unique location to entertain and market worldwide. Yet the military initially chose this once-remote place near Alamogordo, New Mexico, for missile testing and firing because of the isolation and great, empty surrounding spaces.

The land that was claimed from some of the surrounding ranchers did spawn lawsuits, some of which are still ongoing. Rightly or wrongly, White Sands is where America adapted the German V-2 rocket to, at the time, desperate Cold War needs. Charles Leavell was a part of that history. As the decades and adventures revealed in this account, he seemed destined to be near and involved in history, either by making it or by crossing its inexorable pathways.

When Charles Leavell first moved onto the White Sands construction spot, he was really challenged—they had to drill a water well before anything else could start. It was just plain, undeveloped, high desert ranchland, seeming to go on lonely and forlorn forever. However, this vast acreage had been hard won indeed by the pioneer Cox family. Some 109 years earlier, W. W. Cox had felt that a family feud near San Antonio, Texas, called for a change of scenery, so he chose this desolate spot in New Mexico. He dug a cave into a caliche hill and started raising goats. Cox's goat raising dismayed Sheriff Pat Garret, who was a successful cattleman nearby. The Cox family lived there two or three years before moving into a house—a house with walls three feet thick for protection against hostile bandits and Indians. They had the only fresh springwater between El Paso and Tularosa; and whoever controlled the water controlled the land back in those pioneer days.

Radar technical facilities, White Sands Missile Range, New Mexico.

Eventually W. W. was able to increase the Cox family's acreage by buying up homesteads, railroad lands and squatters' rights. They had also converted to raising cattle before 1893 when W. W. Cox bought the San Augustin Ranch. W. W.'s son, Jim, took over the ranch in 1926 when he was thirty-one years old. He soon began to run several thousand head of cattle and many horses. In the mid-1970s Jim's son, Rob Cox, and his wife, Murnie, returned to the old home place. Rob, a decorated WWII veteran himself, said, "Back in 1945, Dad sure hated to sell his land to the government, after all, the family had been here a long time, but with the war going on, he was convinced it was for the good of the country." Jim Cox had sold 90 percent of 105,000 acres to the government.

Indeed it was important. It simply changed the history of the world. The American government wanted the Cox land, along with scores of other, smaller ranches, for a site on which to explode the first atomic bomb, among other uses. Hillsboro, New Mexico, rancher Jimmy R. Bason, of the F-Cross outfit—he was mentored into a cowboy/rancher life as a kid by the Cox family—said, "I don't know how Jim Cox gave it up after all the blizzards, droughts, outlaws and countless other challenges the family had fought through to get it. That's what you call real patriotism."

Bason was helped by the Coxes in acquiring the F-Cross. It joined one of the most famous ranches in the West—the Ladder—on the south. (The Ladder is now owned by Ted Turner and Jane Fonda.) In another of history's curious interweavings, Bason also worked at White Sands from 1952 to 1955.

But here was Charles Leavell stepping into the rapidly spreading puddle of history again. The U.S. shipped ninety-six German V-2 rockets to White Sands after the war. Charles built the first launching site, and the very first one was test-fired from the land where stagecoach drivers, U.S. Mail riders, the contradictory Pat Garret, the great Apache chief Victoria, and countless other famous names of the West had passed through. It arced off the Leavell-built launching pad into the vast, bluest-of-blue southern New Mexico sky to lead off the first step of the great missile crises and standoffs of the Cold War and to humanity's first moon landing. Thirty years later, Charles Leavell's

Titan I missile facilities, Rapid City, South Dakota.

company would construct the almost impossibly huge and precise launching pad for the first Titan missile.

Charles's jobs at White Sands were eclectic indeed, but no less important to the nation and the free world. Not only did his company build the million-dollar rocket testing facilities, including the V-2 rocket and Zeus missile test sites, but their work also included such diverse projects as the movie theater, base chapel, dental clinic and elementary school. The work schedules on the contracts were agonizing, with seven-day weeks and ten-hour days under the tightest security and secrecy possible.

The dedication and never-faltering handling of workloads gave the government tremendous confidence in the Leavell Company. Their cost estimates were found to be eminently trustworthy and were sometimes sufficient in themselves for award of a contract.

On one occasion the Leavell Company was about to suffer a minor loss because of differences of opinion. Harold Tannery, a double-tough job superintendent, got into an argument of some heat with a government inspector. They moved behind the electricians' shed to settle up. Tannery emerged from behind the shed with a bleeding nose and rapidly blackening eyes, saying, "He made a negotiator out of me."

By 1949 the Leavell Company was ready to spread its growing knowledge. Charles negotiated a deal with the huge Utah Construction Company for a joint-venture project at world-famed, world-influencing Los Alamos. They bid $12.1 million on the job and won. Among other complex structures they built there were the Van de Graf Building, the Lab D building, and other extremely complicated environmental controls and disposal systems. The work, like that at White Sands, was top secret and high security. In fact, whenever Charles would leave "the Hill," as it was called, his briefcase was lock-chained on his arm. They knew what they were building but had no idea what the structures would be used for or why. They did know that everything had to be perfection or great disasters could occur.

Before they could finish the project, the Korean War started and critical materials became scarce. Tom Mackey of Utah Construction, along with a brilliant young finance man, Frank Hall, were able to finish the job at no loss. A very close, high-stakes call, but a professional enhancement nonetheless. This gave Charles even more education in the variables of high finance. He would need and use all this accumulated knowledge and experience to move on to multiple, even bigger projects.

After the Los Alamos projects, he entered the field of government housing, incorporating Texas Homes. He and a fellow El Pasoan, Dan Ponder, joint-ventured the first Wherry housing built in the U.S.A. The pressing post-WWII need for housing for Fort Bliss military personnel and their families was won with a $7.6 million bid. Texas Homes grew so much that later, in Charles's middle years, the company had changed exclusively to land and commercial development. With Charles's beloved and brilliant wife handling part of the business, as well as overseeing the auspicious growth and education of Mary Lee and Pete, the Leavells were making lots of money and were gaining the friends and worldly experiences that made up their grand, but risky and obviously adventurous, life.

Near the end of their early years' achievements, Charles developed a very bad backache as a result of destructive wear on the damaged bone in the leg that had burdened him since childhood. Charles, who was still in his thirties, tried to ignore the excruciating bouts of pain, but even one whose demeanor contains alloyed steel sometimes has to seek help. Anyway, even if he could have endured the pain, he could not abide his golf handicap going from eight to fourteen, then to eighteen—that was absolutely unacceptable! He and Shirley talked it over and she was enormously relieved when he decided to seek help at the renowned Mayo Clinic, from the no-less-noted Dr. Mark Coventry. The decision was made to replace his damaged hip with a steel plate and then he was to consider retiring . . . maybe.

The cast covered Charles from ankle to armpit. There is no use dwelling on something everyone can understand. In spite of his attitude of good cheer, this man of such ambition, with a true sense of accomplishment and unbounded courage, must have suffered many kinds of physical hell, and even more mentally. Nevertheless, the operation was successful, and all considerations of closing his construction business and moving into more easily handled investments vanished like a pebble in a vortex. Charles's family, friends, employees and associates were mightily pleased. The 4UR Ranch dream of paradise was one solid step closer for them all.

CHAPTER 8

General Palmer: Civil War Hero and Owner of the 4UR

General William Jackson Palmer was a Civil War hero; a great builder of railroads, towns and mines; teetotaler who threw lavish parties for the elite of the world; and a strange sort of philanthropist. Quite a lot has been written about him, but he has never made the history books one-tenth as much as he deserved. He was a contradictory, brave and dedicated man, as it seems everyone connected to the 4UR Ranch is. He was the first, after the Utes, to really capitalize on the hot medicinal waters and the spectacular landscape surrounding the 4UR and certainly the first to fully realize that it would make an ideal guest ranch. But the first marvel is at how unlikely it was that his trail would cross those of Fremont, Onate, de Anza and the many, many other notable adventurers. Later Charles Leavell rode and walked in many of the same trails, in part of the 4UR's amazing interwoven trek.

Palmer wrote his best friend in 1859:

> Man has to go to the mountains for health and he must go here likewise if he would get a true insight into things. There is a refraction in air of cities and lowlands, like that one meets with on the deserts or in the equatorial seas, when a long coast line or a city with steeples and turrets loom out of the horizon to vanish the next day into vapour.

The twenty-three-year-old Palmer was rapturizing about the Allegheny Mountains of Pennsylvania at this time. His love and

64

General William J. Palmer. (Courtesy, Western History Department, Penrose Public Library, Colorado Springs, Colorado)

reverence for the mountains would multiply many times during his future years of offbeat adventures, on the twining trail of earlier great pioneers, on his way to becoming owner of what is now the 4UR. He would steadily, quietly become a western legend.

Palmer was born in Delaware but grew up in Philadelphia. In a time of manifest destiny and seemingly limitless horizons that enticed and beguiled empire builders, he not only surpassed his own ambitions, but did so through an oddly enigmatic existence. Although a Quaker who worked for nonviolence, he had an uncommon talent for battle and became a Civil War hero. He enjoyed the company of common men with all their foibles yet retained a Victorian formality rare in a man of the West.

The enormity of the contradictions in this man's nature is hard to understand even through examples. How he could build a thousand miles of railroad through the Rocky Mountains, dislodging or blasting away any interference to his steel trail, yet order roads to be resurveyed, at great expense, to avoid disturbing a nesting bird? He made a vast fortune and then, without hesitation, gave away a large portion of it. He forcefully and deliberately sought to accomplish newsmaking deeds, yet he shunned publicity like a surgeon's knife. He dined and celebrated with the rich and famous of the quickly vanishing nineteenth-century world, but in later years found his greatest pleasures playing with children, his large pack of dogs, and riding almost anywhere, high or low, in his beloved Rocky Mountains of the 4UR realm.

In a broad sense, for good or ill, General Palmer stands as a representative of the United States at the time. The dawning of the industrial age found him prepared to conquer it with a background of solid education, strict Quaker values, unflagging optimism and a work ethic that has been carried right on through by everyone, then and now, associated with the 4UR.

Palmer had what all great persons have—vision. The three main Colorado cities that most influence the 4UR realm were built by General Palmer: Colorado Springs, Alamosa and Durango.

The now-booming city of Colorado Springs, near the foot of Pikes Peak, was built precisely to his taste and specifications. It was to be a cultured city of affluent people, great tree-shaded streets, fine universities, hospitals and the most elegant hotels in Colorado. It was and in part it still is. This monumental achievement was accomplished just five years after the city's site was nothing but a handful of roughly constructed log cabins along Monument Creek.

Possibly a recounting of his youth will give at least some insight into this atypical man. By age seventeen, Palmer had obtained a job with the engineering corps (as Fremont had) of the Hempfield Railroad in the Pennsylvania Mountains. This type of apprenticeship was how most journeymen engineers acquired their education in those days. He tenaciously grasped the opportunity.

Fortune was kind indeed. His co-worker, teacher and mentor was the distinguished engineer, Charles Elliot, who designed the trestle across Niagara. Palmer literally slaved from dawn to dark laying out straight paths for the smoke-belching trains soon to come.

General Palmer on iron horse, looking west toward Pikes Peak, the top of his domain. (Courtesy, Western History Department, Penrose Public Library, Colorado Springs, Colorado)

Two years later, Palmer borrowed money from relatives and sailed to England to secure a more sophisticated education in both railroading and mining. Carrying letters of introduction from his former esteemed employer, he stepped off the ship at Liverpool and for months walked from one end of England to the other, staying with Quakers, as much as possible, to economize. After touring the underground hells of the Welsh mines, he came to believe that no one on earth should have to work under those conditions. Later, even when over a thousand men were on his payroll, not one ever had to suffer such indignities.

Palmer visited London and Paris. He listened to lectures by England's top engineers, who sniffed at the young American's eager questions. He considered them pompous jerks in contrast to the

67

working Brits, especially the Scots and Welsh, whom he grew to love. He courageously allowed several of his stories for the *America's Mining Journal* to be reprinted in England They created a stir by introducing many concepts the contemptuous British engineers thought too radical to consider, such as changing from burning bituminous coal in steam engines to the harder and hotter anthracite. Shortly afterward the British began burning anthracite.

On his return to the United States, Palmer undertook many experiments; writing that the U.S. railroads should, in order to get a lot more mileage out of a dollar's worth of fuel, change from burning wood to using coal. Soon the nation's railroads were changing to coal, wherever feasible, and in large part this change was instigated by Palmer's strong words in trade journals.

He left the West Moreland Coal Company before he was twenty years old, going to work for his friend, J. Edgar Thomson, president of the Pennsylvania Railroad. One young executive there who helped Palmer was Andrew Carnegie. Another was Thomas A. Scott, soon to be assistant Secretary of War under Lincoln.

Palmer had gained vast international knowledge and connections at a very early age. In 1859 he arranged for abolitionists to lecture in Philadelphia, even while the nation seethed toward war. His reputation as a leader and an abolitionist spread, and his followers were educated and dedicated. He did not sign up for the Union Army, but rather, with his personal money, formed his own elite cavalry unit, consisting, as he said, of "men of breeding, manners and dash." Even so, his dual (or multiple) nature caused him considerable agony of the soul at this time, as evidenced by what he wrote years after the war:

> While I believe war to be inconsistent with the teachings and example of Jesus Christ, and therefore wrong yet I know that it would have been wrong for me to have refrained from becoming a soldier under the circumstances as they presented themselves in this country in 1861. If it be asked, how I reconcile these conflicting principles, I reply that I cannot reconcile them, anymore than I can reconcile the opposing mysteries of free will and fate.

In answering a question about the company he was putting together, he wrote thusly:

[O]ur enemy have such a large proportion of the better class of men in their rank and file (those who sacrifice something to go to war, who have everything to gain or lose by the result, and who submit with cheerfulness to every hardship and the strictest discipline for the sake of their cause), that the enthusiasm of the whole is thereby raised and in morale their army is as yet superior to ours. That we cannot reasonably expect to defeat this class of men except by sending the same class from the north.

He explained to all who would hear that in his company, "It will be as much an honour to be a private . . . as an officer. While the members would be gentlemen," Palmer wrote, "they would be of the kind who would feel proud to submit to strict discipline, drill and hardship."

The command was to go to a regular cavalry officer and the other officers were to be elected by a vote of the men. Palmer referred to his men as "a picked body of light cavalry, composed of young men of respectability, chosen for intelligence and patriotic spirit, and pledged not to touch intoxicating liquor during their service." After the word got out, Palmer had his men in a few days. He was elected captain. They trained in Pennsylvania, and on December 2, 1861, they left for the West.

Picking up horses in Louisville, Palmer drilled his men rigorously. Here his natural predilection for leadership began to show itself. He was just twenty-five years old at this time, but was already a success in the business world, a leader in Philadelphia's abolitionist movement and an innovator in railroad technology. The Civil War created many such youthful leaders, but for most the war always remained the highlight of their lives. For Palmer, it was only a beginning.

He wrote in a letter home,

It is true that I put the boys through the drill without much mercy . . . but it don't hurt them, although some of them complain . . . more to get the sympathy of their dear aunties and sisters and sweethearts at home, I judge, than for any other cause. The McClellan saddle is preverbially [sic] adapted to the seat of the genus homo and an excellent authority has said, you know, that 'the outside of a horse is good for the inside of a man.' I tell them all that they will thank me for it after the first battle . . . and so will their lady sympathizers.

After more drill, the company rode south to Nashville, and then to Corinth, Mississippi, as a backup to Grant's troops. A few of the men in the company, Palmer among them, saw action at Shiloh, a battle that killed 24,000 men. He bravely led the charge into the midst of manmade thunder and blood, even though the sight of body parts and souls being ripped apart offended his sensitive Quaker nature. He fought and inspired his men valiantly as numberless bullets passed near him. If a single one had caught him solidly, the beginning development of the 4UR would have been different—if it happened at all—and this chronicle would have no reason to exist.

Under the command of a General Buell, Palmer's former company was enlarged to battalion strength. Palmer went back to Pennsylvania to recruit the necessary men—of the same caliber as he had in the company already. In a few days, he raised four hundred hand-picked men. When he asked permission to recruit a regiment, it was granted, and Palmer enlisted twelve hundred men in ten days. The regiment got a new name at this point: the Fifteenth Pennsylvania Cavalry.

Soon after, with most of the new men still largely untrained, fate sent Palmer a rare opportunity. General Longstreet's Confederate forces were marching north toward Pennsylvania, and the only Union force around to stop them was Palmer's recruits. He took two hundred of the men "who knew how to ride a horse" and rode down to Hagerstown, Maryland.

Before confronting an overwhelming force, Palmer slipped into civilian clothes and rode into Hagerstown to have a look at Longstreet's activities. Amazingly, he actually ate with the Confederate officers, thus also learning valuable information about other Confederate troop movements. He then made his way back to his men and wired the information to his commander, General McClure.

The Confederate forces turned back before entering Pennsylvania, not realizing that only Palmer and his two hundred recruits stood between them and conquest of the whole state. Palmer and his men rode into Hagerstown and captured a number of stragglers. His priceless information led McClellan to fight the battle of Antietam Creek, which some historian say could have been the end of Lee's army if McClellan had pressed his advantage.

Having acted as a spy once with favorable results, Palmer talked McClellan into letting him try it again in Virginia. Risking being shot as a spy, Palmer crossed the Potomac River into Shepardstown to see if there were signs of the Confederate Army retreating. He was hidden by a sympathetic family, during which time he heard the enemy riding past the little settlement by the hundreds. A Negro woman servant got him civilian clothes and hid him in the loft as the enemy occasionally stopped to inspect the house. Finally, after numerous close calls, they found Palmer and took him prisoner.

He claimed to be a mining engineer named Peters, but was sent to prison with other suspected spies. He was sent to Richmond by a commanding officer who did not believe his story and sent along a report suggesting that the authorities hang him at once. But the report was lost in a bureaucratic shuffle in Richmond, and Palmer spent the next three months in "Castle Thunder," an old tobacco warehouse that had been turned into a prison—a filthy, rat-infested horror.

Among the three hundred prisoners were twenty or thirty spies like himself. One of them recognized Palmer, told him that the Confederates were planning to send the ironclad *Merrimac* down the James River, and asked him, if he should get free, to get the news to Grant. Palmer, still going as "Peters," arranged to be swapped for a civilian held by Union forces and was released because the Confederates had been unable to prove that he was a spy. He got the word to Grant, the Union Navy launched the *Monitor* to counter the *Merrimac*, and the first battle of ironclads took place.

In February 1863, Palmer returned to the Fifteenth Pennsylvania Cavalry, only to find dissension. A number of the unit's officers rebelled when they found themselves on routine war duty and not, as Palmer had promised, "being kept for special duty." They were put under house arrest until Palmer could deal with them, which he did swiftly.

Palmer, now a colonel at the age of twenty-six, was given an ultimatum by the thirteen rebellious officers. They said they would remain as acting officers no longer because their commissions had not arrived from the governor of Pennsylvania and they were tired of waiting. Most of these men were Palmer's close friends, and they asked him to intercede for them with the governor. Palmer, in what had to be a tough personal decision, told them simply that he could not, because they had already quit, and he appointed thirteen new officers picked from among the men of the regiment. He sent the former officers back to Pennsylvania stunned, and the regiment took on a new, more positive, attitude.

In July, the special regiment fought skirmishes at Chattanooga. Later, the regiment fought at Chickamauga and Lookout Mountain, but it was their semi-guerrilla tactics of night fighting, at places like Saquatchie Valley, Dandridge, Mosier's Mill, Mitchell's Creek and Mossy Creek, and living off the land and covering tiring miles through wet mountain hells that earned the regiment the nickname "Palmer's Owls." Palmer, with two hundred men, chased and captured General Vance and fifty of his men in January 1864. In the fall of that year, Palmer cut off the Confederate retreat at the Battle of Nashville, capturing Hood's pontoon train, two hundred wagons and all the animals. After destroying this prize, Palmer's men captured another supply train of one hundred and ten wagons and five hundred mules. In all, they had ridden and fought four hundred miles in seventeen days—a feat of remarkable dedication and endurance.

After his capture of Vance, Palmer's name was submitted for promotion. After Palmer destroyed Hood's supplies later that year, Lincoln promoted him to brigadier general. This gave him command of not only his own Fifteenth Pennsylvania, but also of the Tenth Michigan, the First Brigade of Gillen's Division and the Twelfth Ohio Cavalry. He was twenty-nine years old.

In April 1865, Lee surrendered, but Palmer's job was not yet finished. Jefferson Davis, president of the Confederate States, refused to surrender. Palmer was ordered by Secretary of War Stanton to "continue the war, live off the country, and pursue Jeff Davis to the ends of the earth if necessary and never give him up." This he and his ardent soldiers did without question or delay.

Charles's grandmother and grandfather, a Civil War veteran, 1910.

They rode southwest between thirty and fifty miles a day in pursuit of Davis, who was with Joe Johnston's army and what was left of the Confederacy's treasure. For this, the final expedition of the Civil War, Palmer commanded even more troops, including nine more regiments. Palmer, leading the Fifteenth Pennsylvania, chased Davis into a trap set by another of his regiments, the Tenth Michigan, which captured the former Confederate president on May 18th. Palmer's men captured Davis's wagon trains and treasure, including $188,500 in gold and millions in Confederate money and bank securities that had been looted during Davis's flight. Every dollar was returned to its rightful owner.

William Jackson Palmer was in a prime position to take advantage of fame at this moment. Many men would have done so. Some successful generals looked to politics, banking, or the writing of memoirs, but for Palmer it was the time to look toward the West. The eternal American West . . . surely one of the great experiences and experiments in the history of humankind in its combination of creation and destruction. General William Jackson Palmer was a huge part of it, a part that led inexorably to the 4UR realm.

Both of Charles H. Leavell's grandfathers fought in the Civil War on the Confederate side. His grandfather Leavell served under General Hood. His maternal grandfather, W. H. Walton, was in the thick of the battle of Shiloh. On the second day he was wounded in the leg but continued fighting. The intertwining of trails, crossings and meetings of this invisible web of history is such that Palmer, or one of his men, could have wounded Charles's grandfather. The meshing of the paths of all these historical folk, who led and were led to the 4UR, seems far beyond plain circumstance; puzzling to some, maybe, but enlightening and awesome to others.

CHAPTER 9

The Phipps Clan of Colorado: Doers and Givers Supreme

Three generations of the Phipps family were a major part of the development of Denver and large parts of Colorado. The Phipps clan of Denver became a part of the 4UR realm and for a brief moment a centerpiece of Charles Leavell's greatest dream.

Denver, the mile-high city, is the center of the Rocky Mountain empire, with all its mining, oil, ranching, tourist attractions and western lore. It is a city built from the raw production, exploitation and often restoration of our great earth. It is now a hybrid, of high technology, old ghosts and wild and raunchy history combined with an appreciation for and love of the arts and sports. Denver is a consummate contradiction. The Phipps certainly did their share to add to all of this, creating both resentment and admiration of their methods of operation.

Senator Lawrence Cowle Phipps arrived on this earth in Amwell Township, Pennsylvania, on August 30, 1862, son of William H. Phipps. His mother, Angus McCall Phipps, was a Scot from the McGregor clan. His grandfather, Henry Phipps, was a shoemaker who emigrated from England in 1832 and hired a widow to help him with piecework on shoes. Henry's sons, William and Henry, Jr., became close friends with the widow's sons, Tom and Andrew Carnegie.

Henry Phipps, Jr. later went into the steel manufacturing business with Andrew Carnegie—the same Carnegie who encouraged and helped young General Palmer (who later owned the 4UR) when he went to work for the Pennsylvania Railroad. Henry Phipps, Jr.'s grandson, Allan Phipps, would one day own the 4UR Ranch and talk hard-nosed dealing on it with Charles H. Leavell.

Lawrence Cowle Phipps went to work, after finishing high school at age sixteen, as a night clerk at a Carnegie Corporation steel mill. Even though his uncle, Henry Phipps, Jr., was Andrew Carnegie's partner, Lawrence started at the bottom (as all our 4UR people would) and was expected to work hard for advancement. He did. Before long Lawrence Phipps was hand-picked, along with twenty-two other young men, for special training—much the same as today's executive management training programs. Carnegie himself started out as a poor immigrant from Scotland. He had begun his working career as a mere bobbin boy in a cotton mill for $1.20 a week, and he, too, had then learned to be a telegraph operator for the Pullman Company. Lawrence Phipps charged up the corporate ladder in this order: assistant bookkeeper, treasurer, vice president, partner and, at the age of thirty-eight, a director in the huge Carnegie Company.

Later Lawrence Phipps would make light of his strong and skillful achievements by saying that Carnegie had launched him upward because of his beautiful specimen handwriting. That may not seem like much today, but it was absolutely vital in those days of handwritten business correspondence. Carnegie was also pleased that Lawrence had learned Morse code.

Lawrence Phipps married Ibrealla Hill Loomis and during their rich marriage she bore two children, Lawrence, Jr. and Emma. Ibrealla died of tuberculosis soon after, and Phipps did not remarry for nine years. He was thirty-five and a senior vice president at Carnegie when he married eighteen-year-old Genevieve Chandler, a renowned Philadelphia belle. She was a great beauty with auburn hair and a wonderful complexion. She bore Lawrence two daughters, Dorothy and Helen.

Then a corporate action occurred that indirectly led Phipps to Denver: the Carnegie Corporation merged with J.P. Morgan's U.S. Steel Company. J.P Morgan was considered one of the most powerful men in America. During one great U.S. financial panic of 1893, he single-handedly saved its railroad industry and thus the lifeblood of the country. J.P. Morgan also encouraged and partly financed Edward S.

Curtis, the famed photographer of the West. In spite of contradictory feelings, mainly because of the lack of full, truthful research, Curtis gave up decades of his life, his health, his family and his fortune in an attempt to save the culture of the American Indians, which by then had been driven almost to extinction. The Carnegie family also made many crucial purchases of photo print sets that greatly helped Curtis in his magnificent obsession. Though it is somehow overlooked by most Indians and white historians to date, this great gesture of life significance was made and helped along by Phipps's most intimate business associates, the Carnegies.

Because Lawrence Phipps had invested all his work and savings in Carnegie stock, he was now worth $30 million and idle for the first time in his life. On a visit to see his sister Fanny, who had moved to Denver for her tuberculosis, he was deeply impressed with Denver's vitality and potential, so he moved his wife, son and three daughters there in 1901. There Lawrence Phipps was eagerly accepted into the elite fraternity of Colorado's upper crust. He mixed socially and in business with the powerful men who fraternized at the Denver Club, such as William Gray Evans, Gerald Hughes, and the Boettchers.

It was Phipps, along with Evans and Hughes, who saved the long-plagued Moffat tunnel and put Denver on a direct transcontinental railroad. The project's problems had cost Moffat his fortune and his life, but Phipps's group carried on Moffat's dream. In spite of much criticism of some of their methods, that was certainly not all that Phipps gave to his adopted state. He spent millions in 1903 to build the Agnes (Phipps) Memorial Sanitorium, in fond memory of his mother. He gave many thousands to the Red Cross and one hundred thousand to help rebuild Children's Hospital. (In later years, Charles Leavell was a member of the board of this same hospital.) Lawrence, like everyone connected with the 4UR, was a self-maker and a giver, not just a taker. There are many contradictions in his life, but the number of his charitable deeds is so great, and so many of them were anonymous, that they are absolutely beyond tracing.

When Lawrence's wife, Genevieve, left him, taking their two baby daughters to New York City, the scandal titillated and obsessed Denver. Lawrence followed her and managed to take back the children.

He named Dr. Thomas Gallagher, the director of Agnes Memorial, as one of the co-respondents in a divorce suit. Carnegie and Morgan demanded that he calm down and settle his domestic problem with civility, in order to maintain control over $10 million worth of U.S. Steel stock. The two giants feared a takeover by John D. Rockefeller. Their advice apparently worked, because Lawrence Phipps did calm down properly. Although he often gave it away, the habit of acquiring and keeping wealth still ruled Lawrence.

Seven years later, he joined a politically powerful family by marrying Margaret Platt Rogers. He was forty-nine; she was twenty-three. Nevertheless, the marriage worked until he died forty-seven years later. Margaret bore him two sons, Gerald and Allan, who were destined to become Denver and Colorado heroes—ironically, considering the Phipps family's many more important gifts—because of a football team.

Along with Anne Evans and a few others, Margaret became Denver's leading patroness of the arts. From casual consideration Margaret might have appeared to be a dilettante social hostess, but her commitment to the arts was real. For the times, she had an advanced and even amazing social conscience, helping young talented musicians regardless of race. She helped educate a young American Indian vocalist and saw to it that a black pianist got a proper first recital. Lawrence Phipps III recalled that Margaret "would give her total dedication and respect to her guests of honor who were blacks, Indians, whatever. If other guests were uncomfortable with this, she said, 'Get used to it. This is the future.'" She helped found the Denver Symphony Orchestra and contributed both time and money to the Central City Opera House Association.

Audree Ducey, after interviewing Lawrence III, quoted him as saying that his step-grandmother Margaret, was "disarming. She could make people feel happy and active. She was a kick in the ass . . . a fun woman who was fun to be with." She died in 1968 at age eighty, having already donated the large Belcaro estate to Denver University. Lawrence III went on, "She had a rich life. Her motive in giving was completely honest and sincere. She was true aristocracy." These are indeed words that deserve attention by all so-called aristocrats.

Backed by powerful figures such as Evans and Hughes, Lawrence Phipps was elected to the United States Senate in 1918 and again in 1924. He applied the efficiency of big business to government and was able to increase needed development in the West while (unbelievable today) decreasing overall federal spending. He won money for national parks, irrigation and land reclamation projects. He took huge losses, of course, in the 1929 crash, but built the palatial, $500,000, fifty-five-room Belcaro mansion at 3400 Belcaro Drive. He finished out his last years as a top Denver philanthropist. In 1940 he gave the Natural History Museum the Phipps Auditorium and then established the Phipps Foundation, which subsequently donated millions to nonprofit organizations. He passed on in 1958 at the age of ninety-five.

He was rumored to have been a candidate for the KKK, and was accused of defending the rich against poor, working people. Maybe so, but his activities supplied large numbers of jobs and services to many. Flip a coin on the late senator.

Considering all the criticisms and jealousies that always plague men and women of great achievement, Lawrence Phipps's life was one of very hard work to make his fortune. His government service was highly commendable. Among his many achievements was his insistence that the mighty Boulder dam's power be leased through private companies instead of public utilities. He also pushed for better roads. He was a personal friend of President Harding and was also the richest man in the Senate (which didn't hurt his power). Lawrence and Margaret's tracks would be hard to follow, but considering the changing times and the changes yet to come, they did a really fine job.

The senator's son, Lawrence, Jr. (known as Lawrie) like his father was a true force in furthering the work on the critical Moffat tunnel and railroad project. While the senator was in Washington, Lawrence Phipps, Jr. looked after the family and its charity work and handled their investments with industry and wisdom. In 1912 he married Gladys Hart, producing three daughters. They divorced in 1930.

Two years later Lawrie married Bertha Richmond. She was twenty-five years younger than he. They had three children, including Lawrence III. They divorced in 1941. She was awarded custody of the children and moved to a ranch near Sheridan, Wyoming, where she remarried.

But this is, in the end, the story of a special dream and place, so we must delve into the life and times of the Senator and Margaret's two sons, Gerald and Allan—the latter, of course, would own the 4UR. While their father was in Washington, they also went east, attending private schools in New Jersey. Afterward they went to Williams College, a noted private school in Williamstown, Massachusetts. Allan furthered his own education at Oxford in England, returning for law school at University of Denver. In total, he acquired three degrees in jurisprudence. Gerald graduated cum laude at Williams with a bachelor of arts degree. Two years later, in 1938, he was admitted to the Colorado Bar.

Following the tradition of starting low and working up, Gerald became a clerk for operations at the Denver and Rio Grande Railroad Company in Denver, a company founded by another of the 4UR realm alumni. Yet again, the criss-crossing trails of fate lead ever nearer to Charles Leavell's beautiful dream.

Having earned a secure job, Gerald married Janet Alice Smith on July 24, 1937. Her father, Herbert, was the president of U.S. Rubber Company in New York. Allan married British-born Doreen Evans. They had three children.

When Lawrence Cowle Phipps died in Santa Monica on March 1st, 1958 at the age of ninety-five, he left an estate of over $10 million. The generous Phipps Foundation was handed over to a board of trustees that included Allan. But here's the twist of circumstance that leads us to the crux of the meeting between Allan Phipps and Charles H. Leavell: in the will, Allan was left the Wagon Wheel Gap Ranch near Creede, Colorado, now known as the 4UR.

Both Gerald and Allan were in national societies such as Chi Psi and Phi Delta Phi. They had interests in banking and served on many boards of directors. Allan spent enormous amounts of time with arts and nonprofit groups, often as president or trustee, including the Denver Symphony Society, the Denver Museum of Natural History, St. Luke's Hospital and the Winter Park Recreation Association. Gerald was a member of so many city and state organizations (including both the U. S. and Denver Chambers of Commerce) that it is no use attempting to name them all.

All their charitable work wasn't what made them Denver heroes, though. Gerald loved sports, playing baseball as a youngster, tennis in college, and golf ever since, so it does not seem strange that he would sit right on top of Denver's sports empire. He became a director for Rocky Mountain Empire Sports Corporation and owner of the Denver Bears baseball team. During the 1947 season, Gerald's construction company was contracted for repair work on Merchant's Park, home of the Bears. Instead of taking pay, Gerald took stock for the repairs and made the same move when he contracted to build the Bears stadium for a 1960 opening. When funds ran out, Gerald took even more stock.

In the meantime, Allan was creating business of his own in the Belcaro Investment Company and the Highlander Hotel Corporations, but he suffered changes in his personal life as well. In March 1954, he was divorced from Doreen, which resulted in a substantial settlement and alimony and child support payments. He soon met and married Clara Van Schaack Combs. They were wed in 1955 and later had a son whom they named Lawrence Cowle, after his grandfather.

By 1961, Empire Sports, which owned the Denver Broncos, was in financial trouble. Several firms, including one in San Antonio, wanted to buy the Broncos, but the Phipps wanted the team to stay in Denver. Gerald moved swiftly. He made it possible for Allan to buy stock and then he formed a syndicate of other stockholders.

Unlike most Phipps enterprises, the Broncos were a losing team during their first five years and had accrued a loss of $2.2 million. The Phipps brothers by then held 42 percent of the stock; the syndicate held 52 percent and the other 6 percent was owned by individuals. The Phipps brothers' total cash investment only amounted to $150,000, but their intangible contributions were enormous. The team was reportedly up for sale in February 1965. Atlanta-based Cox Broadcasting offered to buy and move the team for $4 million. Empire Sports was hurting financially, but the Phippses and others insisted that the team be kept in Denver. The Phippses said that their share was not for sale. They agreed to buy 25 percent of the syndicate's stock for a mere $1.5 million. To their surprise, they got it, and thus were able to save the team for Denver. Gerald said that his real business was now construction and that the Broncos' leaving would have had a negative effect on all of

Denver. He thanked a Mr. Kung and other board members for turning down better offers so they could keep the team in Denver. The Phippses had borrowed $1 million of the money from the First National Bank of Denver. Over the years, the Phipps brothers came to own 99 percent of Empire Sports, Inc.'s, stock (the remaining 1 percent was held by attorney Richard Kitchen).

After having a 4-10 season, just knowing that the Broncos would remain in Denver changed the city to one of the most active sports towns in the AFC. A real football craze blew over the city. Season ticket sales climbed to all-time highs. Banks in the area got involved, making interest-free season ticket loans on installment payments. This unique idea was very successful.

Then, in 1967, trouble entered the happy scene when the Denver Metro Stadium was found to need $20 million in repairs. After all sorts of negotiations and offers, the Metro Stadium bond issue submitted to Denver voters lost by 90,195 to 52,787. The Phippses were shocked and disappointed, feeling now that the Broncos were without a home. They sought new options. Because of the Bears' schedule, the Broncos had to play some preseason games at the University of Denver Stadium. This was also deemed inadequate. They considered renting Folsom in Boulder, but they started their 1967 season in Bear Stadium.

The defeat of the bond issue gave Denver a bad name nationally. Other team owners hesitated to bring their teams there and fans in other cities were not happy to have the Broncos as playing guests. Oddly, while the Broncos played badly in this dismal situation, the number of die-hard fans increased.

Almost one year after the Phippses' offer to sell Bear Stadium to the city, Mayor Currigan made the deal on February 15, 1968, for $1.8 million. The money was to be raised through private funds and Metro Stadiums Inc. Immediately the city issued a $3 million bond to expand the stadium from 30,000 to 51,000 seat capacity. The Broncos finally had their Mile High Stadium in the Mile High City, largely due to the Phipps brothers' persistence.

The new teams coming to Denver gave the Broncos a large passel of rabid fans. After 1969, every Bronco game at Mile High Stadium has been a sellout. The subsequent subsidiary employment and money distributed to Denver and the city's environs has been enormous. The Broncos had their first winning season in 1972 as the growth increased. Even so, there were political fusses and scandals of all kinds.

In particular, local television blackouts caused dissent. Representative Patricia Schroeder fought and voted for lifting local blackouts. Her opponent, Dan Friedman, backed the blackout. Allan Phipps agreed with Friedman, saying, "If you don't pay the freight you don't see the game free. Our belief is that any business has the right to be compensated for the product they sell." Friedman lost the election to Schroeder.

No matter; the Broncos kept improving and had a 12-2 season in 1977, earning a Western Division championship and a trip to the playoffs. Beating both Oakland and Pittsburgh, they won the American League championship and went on to the Super Bowl in New Orleans. After that there were several more championships and an ever-growing number of screaming "Orange" fans. With the success and the Denver "home" assured, the Phippses sold the team to another steel magnate, Edgar Kaiser.

The three-generational Phipps touch was apparent, as they received $30 million after an original total investment of between $3 and $4 million—a nice return. Even with changing times and situations, the Phipps clan's wealth, albeit scattered through estates and other transfers, is still there. Though often in disagreement, the collective family is still very close. The Phipps spouses, children, in-laws and business and social associates form an integral part of Denver and Colorado society. Their genes may carry on forever. Without the three generations of Phippses, Denver and Colorado would have been far different.

It seems a great irony now—or perhaps a touch of destiny playing out—but in 1970 Gerald Phipps was elected to the board of Rio Grande Industries, Inc., founded by General Palmer. Serving on that same board was a Mr. Charles H. Leavell of El Paso, Texas.

All aboard for the Rio Grande Industries Board of Directors.

CHAPTER 10

Explosions of Energy Catapulting Through Canyons of Dreams

By the early 1960s Charles Leavell was steaming ahead and his business interests were expanding like summer storm clouds. Mary Lee and Pete were getting their formal educations, as well as following the tradition of apprenticeship in various Leavell projects. Shirley—besides raising, loving and guiding the children—accompanied and advised Charles on most of his eventual worldwide ventures. They shared the hardships, the winning and a lot of fun activities that always balanced and enhanced their love for their work and their family—including the corporate one. All the adventures created and solidified an indestructible respect and love among the Leavells.

In spite of occasional weariness and any handicaps, Charles was a fine dancer, and he and Shirley both loved this activity. They shared it, all over the globe, privately, in classic clubs or in just plain joints when nowhere else was available.

In spite of what would have been a daunting schedule for most, Shirley flourished and helped launch many charitable campaigns, including her pet one—the one she helped develop in El Paso—the Y.W.C.A., now the largest in the United States. She continues this work today with as little fanfare as possible, while still maintaining her strong and respected directorship positions in the Leavells' chosen remaining companies.

There are far too many key people in the vast Leavell projects to name them without sounding like the phone book of a medium-sized city. However, one of Leavell's El Paso High School classmates, Joe Lea was a key player. In fact, Charles Leavell referred to him as his #1 man. Joe was executive vice president, administrator and secretary for the Leavell companies. He was bluntly honest—a trait Charles still respects anywhere he finds it—but with a grand sense of humor. Joe dealt with every possible type of problem, from the smallest to the largest, with dedication and finesse. He was Charles's second conscience and devil's advocate.

Joe Lea's brother, Tom, a great and totally dedicated artist/ writer, became Charles's closest friend and an occasional celebratory partner on many widespread parts of this earth for a span of over fifty years. Shirley Leavell and Sarah, Tom's wife, completed the quartet of love and laughter that became a major part of this odyssey of dedicated, patriotic, hard-working and achieving Americans. Eventually, their friendship, their accomplishments and their activities of powerful influence and beauty acted more and more as a focal point for the velvet web of fate and history that drew them closer to the 4UR.

The period of most fruitful production for Charles H. Leavell and Company encompassed seventeen years, and involved construction projects from coast to coast in America, Africa, Europe and the Middle East. The business consisted of six major design categories:

1. Business and industry
2. Dams, waterways and bridges
3. Power and utilities
4. Health and education
5. Development and leasing
6. National defense and space programs.

Colonel N. J. Riebe, retired from the U.S. Army Corps of Engineers after supervising the Shumaker Ordinance Depot in Arkansas, joined Charles and was the person who really propelled Leavell into the big time. The project was a joint venture with Utah International Construction Company to construct an ammunition defusing building and storage facilities. At first it looked like this job was headed for a large loss. However, their accountant knew that the effect of their newly forming system had not yet come into action. Charles probably thought he was nuts, according to the accountant, who suggested they buy out the Utah company. No matter—Charles went along and like so much of his life, disaster turned into jobs that were extremely lucrative.

Since Charles Leavell was joining the big boys in construction, one would think he would headquarter in one of the major cities, such as New York, Los Angeles, or somewhere in between. But because of his love for his El-Paso-region background, the people of their birthplace, the beautiful and sumptuous home he and Shirley had built in Red Rock Canyon and their children, they simply preferred living and headquartering in El Paso. Their judgment and choice were obviously correct. The advantages of private airplanes made it all possible. Company pilots flew out of El Paso to all points in varied planes, but a Lockheed Super Ventura—named "The Pregnant Gypsy" for her somewhat ungainly appearance—did a majority of the winging.

The Leavell's Texas Homes Corporation built the $28 million Capehart Units at Fort Bliss, Dyess and Holloman Air Force Bases and William Beaumont Hospital. Charles couldn't borrow that kind of money in El Paso, so he went to Morgan Guaranty Trust Company in New York and "rented" money. His reputation was such that the eastern bankers took bonded contracts as collateral, with no personal guarantee required. As competitors agreed, "That took clout."

Texas Homes, headed by the indomitable Harry Buckley, was a separate entity from the Leavell's other companies and endeavors and it always made money. It was a favorite pup out of a big litter. When it was later sold to Rio Grande Industries (with Charles remaining on the board), it had become the biggest developer in the entire region. By 1970, it had moved from its beginning construction phase into land development that formed a "ring" around El Paso, according to author Bill Lynde. Texas Homes projects included more than three thousand acres in Coronado on the west side, with several partners; Westwood and Eastwood near El Paso Country Club; and Buena Vista.

Paradise Island Hotels, Casino and Villas, Nassau, Bahamas.

Charles took a loss on the Campbell Soup Company project plant, built in 1962-1963 at Paris, Texas, but made up for that at Paradise Island, Nassau, in the Bahamas. They built two hotels, a gambling casino and villas, plus recreational facilities for Resorts International, which had purchased the Paradise Island land from Huntington Hartford, the A & P heir. They also designed and built a high-span bridge and an eighteen-span vehicular highway connecting with downtown Nassau.

In spite of his profits in the Bahamas, Charles decided to depart because a newly elected prime minister insisted that his nephew of little accomplishment be maintained on the payroll at $50,000 a year. Charles was no fool about politics. That was what made wheels—that had been oiled—turn. But there was a limit and an integrity point that must never be breached. The nephew was never hired.

Leavell's companies moved with rapidity and skill right into major business on the West Coast, opening a branch office in El Segundo. One of the diamonds in their ever-expanding tiara was a $13.3 million federal office building and postal facility in West Los Angeles.

Paradise Island Bridge, Nassau, Bahamas.

The edifice was dominating indeed with its seventeen-story tower. Charles Luckman was the designing architect. In rapid sequence came the twelve-story Le Sage building at Vermont and Sixth Street; the $12.9 million International Exhibition Facility in New Orleans; a twenty-one-story bank building in Houston; an entire company town, in a joint-venture with Brennard Construction Company, for Phelps Dodge in Tyrone, New Mexico, site of one of the world's biggest open-pit copper mines; the Grand Hotel near Disneyland in Anaheim, California; and an $8 million downtown office building, plus the Park Plaza apartments and the PNM Building, in Albuquerque, New Mexico.

One New Orleans job was a perfect example of how the best in the world could be defeated by circumstances truly beyond anyone's control, except possibly the great mystery in the sky. It also exemplified what a heart-jarring, massive roll of the dice international contractors take on every major bid.

In New Orleans, they were driving 2,665 feet of pilings into unstable soil, while trying to deal with four nearly insolvent

Park Plaza Apartments, Albuquerque, New Mexico.

subcontractors and eleven different architects and Cajun-style politics that altogether spelled a big loss. They did find one treasure, however, as they dredged the mud from the old Mississippi River building site. It was an eight-foot-long anchor with a hundred-foot-long chain attached. It was believed to have come from the Spanish governor Ulloa's ship when it was cut from its moorings by angry Frenchmen two hundred years earlier.

As always, no matter what difficulties and impossibilities they faced, the Leavell group tried to mix in some fun. They organized a Dixieland band and brought their horns and drums to the job. At lunch break, the Leavell Hardhats' Band cascaded music across the doomed site at the foot of Canal Street in joyous defiance.

Park Central in Denver was another tune altogether. This design/construct, $21 million contract, owned and built for the purpose of tenant lease by the Leavell Company for Leavell Enterprises as the leasing owner, occupies a full square block, with three interconnected high-rise buildings totaling six hundred and ten thousand square feet of floor space. At the start, C.H. Leavell Company owned 52 percent, with Rio Grande Industries and Central Bank holding 24 percent each. This was a racehorse design on a two-year completion schedule. However, the Leavell Group maneuvered one of the top coups in its corporate history by merging three Leavell affiliates—Texas Homes, Frizzell and Company, and Leavell Enterprises—into Rio Grande Industries, a New York Stock Exchange company, and forming a new RGI division known as Leavell Development Company. Both sellers and buyers benefited substantially.

Charles gives enormous credit to Ralph Frizzell, a Leavell vice president, who knew the politics of Denver's business power structure. Frizzell summed up Charles and his companies and his concepts in these revelatory and incisive words: "a people company, a lean machine whose success rested on its family approach, always operating in the client's best interests."

Another perfect diamond in the Leavell tiara was, naturally, a hometown project—a twenty-two-story tower plaza for State National Bank, which had been the initial sponsoring bank for most of Charles's projects. If one has even the slightest interest in a world where most of

Park Central Office and Banking Complex, Denver, Colorado.

the populace is surrounded by some sort of construction projects, from the pyramids to the Taj Mahal to the Empire State building, the El Paso project's unique manner of awarding the contract must be told.

For this multimillion-dollar job, eternal rivals, Leavell and Robert C. McKee, were invited to submit bids. At the opening of the bids, the architects claimed the bids were a tie—a dead heat. Charles said, "Oh no, it's impossible for us to tie. There *must* be a winner." There had been thirty or forty alternatives and everyone knew it. Charles continued, "What are we getting at here—" The architects'

State National Bank, El Paso, Texas.

representative answered that he wanted to get the bids lowered by several hundred thousand dollars and to have the two contractors joint-venture the project. He then left the room and placed the outcome in the hands of Charles Leavell and Dave and John McKee to figure out.

They agreed to joint-venture after Charles insisted that a coin toss decide the sponsoring contractor. His 1881 silver dollar would make the decision. The McKee brothers came right back, demanding that there be two coin tosses, the first to see who would win the right to make the second and deciding call of sponsorship. A bank clerk was called to toss the coin. Leavell won the first flip, calling "heads," and went on to win the second toss with another "heads." To the everlasting credit and spirit of all, the job came in on time and under budget. Of course, Charles had high praise for all concerned.

During this same seventeen-year period of enormous movement of men, material and money, all around and beyond the United States, one of the companies' prime accomplishments was the Homart Shopping Centers, a subsidiary of Sears and Roebuck, all across the United States. Eight of these sophisticated centers were built for $92 million with a minimum of problems. These first-rate, covered malls stand—each anchored by a Sears outlet—in Dallas, Texas; Independence, Missouri; Westminster and Brea, California; Eden Prairie, Burnsville and North St. Paul, Minnesota; and Chicago, Illinois. During this same seventeen-year time of miracles, mixed with an occasional disaster, came the building of several dams, waterways and bridges.

Another very large and bright gem was added to the Leavell construction crown amidst all these massive and often critical structures. It was the $135 million Dworshak Dam and the $45 million power house constructed in a remote area of Orofina, Idaho. It was, at that time, the Corps of Engineers' biggest civil works contract ever awarded. The operation, a joint venture of Dravo Corporation and Leavell, raced around the clock of time twenty-four hours each and every day. Even so, it took six years and 6.6 million cubic yards of concrete—batched on-site to save $7 million in construction costs. This was a project of history-making proportions. It powered nine hydroelectric plants downstream, came in on schedule and was right on the money, and as always Charles gives the credit to the distinguished managers who headed up a five-man management team.

Another type of history was made through a "hell to heaven" project at the Jonesville Lock and Dam in Louisiana, between 1967 and 1970. This $18 million project was a mainstay of the Corps of Engineers' plan to deepen and straighten the Black and Ouachita Rivers to allow ships with nine-foot drafts to safely navigate these water roads north for 350 miles to Camden, Arkansas. The total price was to be $94 million. The project was about average in difficulty, similar to the $14.2 million Arkansas Lock and Dam #1 that they had finished in 1966. However, in 1969 a conflict in funding arose. Money for the Jonesville project in Louisiana was frozen at the halfway point and the Charles H. Leavell Company was given an option of shutting it down or furnishing all the money for completion. They shut it down and allowed the diverted river water to flow back into the sixty-foot-deep, one-hundred-acre lock excavation. It remained so for six months. It was left there with the steelwork sticking up like the beginning of a modern-day Atlantis, or a watery cemetery for Charles and company.

The Leavell Company had filed suit for damages of $1.7 million and lost, but they appealed and won. As Charles naturally would say, "Through brilliant arguments of attorney William A. Derrick, the case set legal precedent, forever changing the law and rewriting the Special Condition Section of Corps of Engineer contracts." This history-altering legal victory earned public recognition for Charles. The *Engineering News Record* honored him as "Man of the Year" with his picture on the cover.

Experienced professionals, Bill Sanders and Bob Hurst, entered the scene as project manager and superintendent, and finished the Jonesville project in style and profit.

These water projects marched on in production like a Marine band. There was the $78 million Amistad (friendship) Dam across the lower Rio Grande for the International Boundary and Water Commission, joint-ventured with one Mexican and two U.S. contractors; a $4.3 million waterway addition to the seaport of Houston; and a 1,920-foot-long natural-gas single-suspension bridge built in the rugged wild country of Wyoming for the El Paso Natural

Amistad Dam, Del Rio, Texas.

Gas Company. This latter project, a steel and concrete structure that crossed scenic Flaming Gorge Reservoir and carried a thirty-four-inch natural gas pipeline, won an award of merit from the American Institute of Steel Construction.

Structures in Leavell's power and utilities categories stretch from the states of Colorado, Arizona, Texas, Colorado, Nebraska and California in the United States to the Sudan in Africa. Later, the often humorous and sometimes threatening facets of the growing adventures in power structures (which included nuclear plants, Titan missile bases and projects in Khartoum, Thailand, Liberia, and China) will find their proper place in this narrative.

At this point, Charles had a $600 million open line of bonded credit. As previously noted, it was a staggering sum then and is nearly unimaginable in today's dollars. By the year 1967, Leavell's international engineering and construction firm company was number six of the top ten largest, in volume of work, in the United States.

During the latter part of these special business years, Charles started buying and selling ranches in both New Mexico and Colorado, making a nice profit each time but really searching for a replacement for his dream of the bountiful, bewitching, trout waters of Goose Creek. The $3,000 loan of faith from his young bride had certainly paid off for both of them, bringing them closer to the modest, but supreme, dream of finishing it all off in pleasure and fishing for the elusive brown at the 4UR.

CHAPTER 11

There Is No Night in Creede, or So They Once Said

Even if Creede were not the nearest town to the 4UR Ranch, and even if General Palmer had not been destined to build a railroad to the town, Creede would still be worth writing volumes about. It seems foolish to dismiss the possibility of another real mining boom of some kind returning to Creede someday—any day. In a way it already has. The streets are filled with more and more tourists each year, absorbing—often in awe, always in wonder—the spirit of the multitudes of miners, con men and women and other pioneers who created Creede. The cleverly built, well-maintained little town winds narrowly between Bachelor and Mammoth Mountains of the San Juans. Evidence of mining is everywhere, in tunnels, waste dumps and preserved mining equipment and relics.

Creede has been both bountifully blessed and tragically cursed throughout its important place in western history. Millionaires were made overnight and business and land empires were established because of its great and abundant mines of silver and other precious and semi-precious metals.

Its great altitude of 8,800 feet afforded cool summers with lots of wild game and endless fishing in its streams. The mountains also gave timber for building and mine exploration, but they were sometimes blasted with a fury of snow and ice more dangerous than the exploding powder in the mines, and people foolish or unfortunate enough to be caught out in the maelstrom often perished. At the center of this contradiction is the fact that the Rio Grande's headwaters are a few miles above Creede; because of the many creeks from the vast drainage area, it swiftly builds into a river. The river is what holds together this unbelievably vast terrain of such majesty—and always will, regardless of fortunes of chance with minerals and travelers.

In the summer of 1889, two prospectors named George L. Smith and Nicholas C. Creede were scouring the mountains near what is now the town of Creede. They had driven their prospector's picks into many promising rocks before Mr. Creede fortuitously tapped into a rich vein of ore. He was experienced in the mineral world and his eyes flew open as he saw the rich silver in the vein. His mouth simultaneously flew open, uttering the words that would change the history of a vast area and thousands of lives. "Holy Moses," he shouted to his partner. This was a mild expletive indeed for such a discovery. They immediately staked out claims and that winter, in Denver, sold the mine for $75,000 to David Moffat. This, of course, is the same David Moffat who would later become famous for the Moffat tunnel that opened up a transcontinental railway near Denver; the same tunnel that our 4UR people, the Phippses, would play such a big part in completing. At that time Moffat was president of the Denver and Rio Grande Railroad, founded by another 4UR owner, General William J. Palmer, who brought that very railroad to this very area and town.

At any rate, Moffat's business reputation was so great that the word of this purchase started the first of many wild and lucrative mining booms. The town, which at one time would reach 10,000 souls scattered along a narrow canyon that in one place is about a hundred feet wide, was first called Willow. The camp was properly renamed Creede in 1890 and incorporated into a city.

Like all these mountain mining towns, it became known as a "ripsnorter." Most of its inhabitants were wilder than outhouse rats. It was without doubt one of the West's wildest, as well as richest, mining camps. Its powdered streets were trod by miners, speculators, bunco artists, saloon keepers, gamblers, whores and thieves of assorted accomplishment. There were others who came later to negate this idea of a totally sinful city. In the beginning, though, graves were being dug twenty-four hours a day, everyday, for as everyone knows there was no night in Creede.

Creede in 1896. (Courtesy, Denver Public Library, Western History Collection; photo by O. T. Davis)

It was the most natural thing imaginable that Nicholas Creede would return to the scene of his first great discovery. In succession he located the Ethyl and then the most storied mine of all, the Amethyst. The latter produced over $2 million worth of ore in the first twelve months of operation. Among the other noted mines were the Last Chance, and nearby Bachelor Hill, Solomon, Texas Girl, Wandering Jew, and New York. Of course, the fact that some of the ore was assaying at $5,000 a ton only created more lust for wealth—lust that often approached hysteria. A veritable frenzy of activity opened many more fabled mines. The Champion hung on a precipice perhaps a half-mile higher than Creede. As always, the unappreciated and overlooked burro and its offshoot, the mule, were used to pack the rich ore over a four-mile switchback trail ever downward, under great stress, to the mills that would turn it into a purity of metal that could purchase any luxury the world afforded.

Imagine the dry, powder-stomped streets or dust halfway to your knees made by the boots of thousands of dream-trackers. And then the belly-deep snow in the winter, often accompanied by winds that could freeze an eagle in flight. The raucous babble of the gambling, the brothels, the liquor dispensers and the hotel restaurants combined with the din of the blacksmiths, the clanging of iron mining equipment, the blasting of powder in the mines and from the barrels of guns, the laughter of winners and the sad sighs of losers. The place was a living entity of capricious connivance, a festering wound of wealth to feed all the greed in Creede.

Hundreds of shacks rose from the ground practically overnight to spawn the town. Some historians have recorded more than five hundred in less than ninety days. Into this earthly vortex poured over three hundred new inhabitants each day. Even among this all-out grab for glory at any cost, any deed, a few desperately struggled to construct permanent houses and attitudes of decency. But before that could take hold, mightier powers than newly acquired wealth would be revealed.

A flash flood washed away a great portion of downtown Creede. Rebuilding for business moved at a furious pace to capture the wealth before it left. A few weeks later, on June 5, 1892, a fire broke out in a building near the center of town. One of those famous cooling breezes, which Creede's present-day tourists and citizens enjoy so much, fanned the fire into a crackling, roaring inferno as it leaped from one pine building to another with terrifying speed. In a very short time it was over. Creede's business district was a blackened hell that flared a few flames here and there, hungry for unburned wooden food, and then settled into smoldering ashes.

Before the fire had completed its ravaging of Creede, the dominant nature of these precious-metal-made monsters leaped—almost as fast as the flames—into action. The *Denver Republican* reported: "Following the fire, a wild debauch was entered into by all the sots and fast women of the camp. Free liquor was had for the stealing, and many cases of wine, bottles of whiskey, cigars, and such goods were seized and hundreds were drunk before the flames half burned down."

Multitudes were left homeless. Business owners of various sorts were broke. Creede was changed forever. But the great miseries were not through yet. Three years later a second angry fire swept through the town—and there were still others later.

Amidst all this accumulation of wealth, and the massive, varied activity of the boom town, migrated some who were notorious enough to make lasting names for themselves. Bat Masterson, tough as new nails, attempted to control the town from his big Watrous saloon. He couldn't quite do it. Poker Alice and her buddy, Calamity Jane, and Soapy Smith, who was one of the greatest of all bunco artists, surely thrived while there. Creede's most famous backshooter, Bob Ford, was even immortalized in a popular song in the 1880s that went:

> Now Jesse had a wife
> Who mourned for his life
> Three children they were brave
> But the dirty little coward
> Who shot Mr. Howard has
> Laid poor Jesse in his grave.

*William Barclay "Bat" Masterson.
(Courtesy, Denver Public Library, Western
History Collection)*

Of course, "the dirty little coward" was Bob Ford. His brother, Charlie Ford, had helped Jesse James and his gang rob a bank. The Ford brothers were aware that Jesse was living in St. Joseph, Missouri, under the pseudonym of Howard. They arranged to have breakfast with Jesse and his trusting family in the Fords' tiny farmhouse. (This structure, complete with bullet holes in the wall, was later moved a few blocks to a popular museum and is today open to visitors.)

As later testimony revealed, the two brothers (ironically) sat on a loveseat. Jesse removed his gun belt, said to hold four revolvers, and laid it on a bed. In an even greater irony, he stood on a chair to dust a picture on the wall. Bob Ford moved instantly, pulling his pistol and shooting Jesse in the back of the head. Jesse fell, smacking the floor hard but dead before the impact.

Unbelievably, Bob Ford turned himself in for the killing of the famed Jesse James and pleaded guilty to murder. He was sentenced to die by hanging, but again, amazingly, he was granted a full pardon by

Poker Alice Tubbs. (Courtesy, Denver Public Library, Western History Collection)

Governor Thomas Crittenden two hours after the sentencing. How could all this happen? Those are the known facts, but historians still disagree on many points. There is some evidence that several supposedly secret meetings had been arranged with the governor shortly before the James murder. The almost-instant pardon naturally lends some credence to this theory. Both Bob and Charlie Ford swore they had killed Jesse in self-defense even though they admitted he was unarmed. Their reasoning was that Jesse would have killed them because of "treason." If he had known of, or even suspected, their deceit, though, it hardly seems likely that he would have unarmed himself. The two Fords were supposed to have split a $20,000 reward from Governor Crittenden, but Bob Ford later told a reporter it had somehow shrunk to $1,000 by the time it reached him.

The Ford brothers left Missouri swiftly and for good reason. The James gang had became extremely popular among the common folk because they robbed banks and railroads—both establishments heartily

Mary "Calamity Jane" Bourke. (Courtesy, Denver Public Library, Western History Collection)

Jefferson Randolph "Soapy" Smith. (Courtesy, Denver Public Library, Western History Collection)

disliked by these people. Charlie Ford's life fades out here, but he did show up at Creede briefly. His brother, Bob, is well documented straight to his violent death. Bob headed to Creede right in the "hot" period of its mining boom. He built a fancy saloon with his share of the reward money, calling it Bob Ford's Saloon. He reveled in the notoriety of being the man who killed the famous Jesse James and made no attempt to hide his deed. He entertained lavishly such aforementioned notables as Bat Masterson, Soapy Smith and Poker Alice, more often than not in riotous revelry.

On June 5, 1892, Bob Ford's saloon mysteriously caught fire. Many thought—then and now—that this was the initial fire that caused the first flaming destruction of Creede. Whatever the truth, in just a few days Creede's decision-making committee of twenty ran Ford out of town for drunkenness and random shooting. Ford was again

Robert Ford. (Courtesy, Denver Public Library,
Western History Collection; photo by N. A. Rose)

"honored" as being the only businessman ever run out of town in the history of Creede.

Ford sent word to the committee promising to behave himself and, not too surprisingly, was allowed to return, accomplishing another first for the strange town of Creede. He built another saloon on the site of his first one. That same year of 1892, a man named O'Kelly shot Ford in the back with a shotgun. Ford, by his charming nature—something like a cross between a wolverine, an angry rattlesnake and a black widow spider—had made for himself more enemies than he could look for over his shoulder. Enough reasons for the killing were brought forth to nearly fill the Grand Canyon. One Creede historian said that if all the people who told him they had seen the killing had actually done so, Ford must have been shot in an athletic stadium.

Some accounts said that killer O'Kelly was a James gang partisan. O'Kelly did get a prison sentence but was pardoned six years later. Two

Ed O'Kelly. (Courtesy, Denver Public Library, Western History Collection)

years after his release, he was shot to death by a law officer in Texas. Ford was buried in Creede in what was described by locals as "a nice part of a cemetery for undesirable characters."

A short time later, Ford's wife had his body exhumed and returned to his native Missouri to be buried there with a simple headstone. Oddly—if anything could be odd about Creede then—she then returned to Creede to work in a dance hall. Her overall intelligence would seem at least equal to that of her husband.

If you had a city of a million living souls and 999,000 of them were good citizens, it would be the other thousand for which the city would always be known. Creede had a better average than that. Probably half its population was made up of God's children or representatives thereof. The duality of the town is perfectly demonstrated by reports that often two or three hundred men and women would stand in the back of a den of sin to listen to a traveling preacher, while from the front part of the establishment emanated the sounds of competing fun and frivolity, as the gambling and the drinking proceeded unbroken. Soapy Smith added to this duality by supporting the local clergy with money and verbal boasting in their favor. He fooled only the fools.

Schools were established and teachers of real dedication to their profession came and taught the 3Rs. Some even attempted to point out the finer things that life had to offer beyond killing, whoring, and thievery. Given this setting, they must have been brave beyond mere dedication!

Fine church buildings were eventually constructed and supported quite well, considering what surrounded them. Then telephones and lighted streets helped the law arrive and survive. There was the Collins Opera House and the Theatre Comique. Newspapers created a voice for those on the other side of town. A. H. Wasson founded a courthouse he had built himself at his rival ranch town, three miles south of Creede, and it was designated the county seat. But now the "good" people of Creede had power, and one dark night they went to Wasson and hauled the county records to their own city. Sometime later they even chopped the courthouse into three pieces and moved it to Creede. It became obvious that one must not fool around with the righteous.

Creede had sort of grown up—in part, at least. One must never forget that the mines may open up again some day, as they always have in the past, and doing so remember what that famous newspaper publisher (*The Chronicle*) and poet Cy Warman, wrote about Creede during its wildest days:

Here's a land where all are equal
Of high or lowly birth—
A land where men make millions
Dug from the dreary earth
Here meek and mild-eyed burros
On mineral mountains feed;
It's day all day in the daytime
And there is no night in Creede.
The cliffs are solid silver,

With wondrous wealth untold
The beds of its running rivers
Are lined with purest gold.
While the world is filled with sorrow
And hearts must break and bleed
It's day all day in the daytime
And there is no night in Creede.

It doesn't take a great thinker to realize that a place with no
night—like Creede—would soon find it a waste of effort to try to hide
anything at all. It seems to have had extra time to make twice as much
history as otherwise possible, and it took full advantage of the
opportunity.

Now that night has returned to Creede, all the great trees on the
mountains, the deer, the elk and the bear have come back home as well.
There are over two hundred miles of fine fishing waters, and the town
itself is a gilded and friendly postcard from the past. The night knows,
however, that prospectors still roam the mountains.

CHAPTER 12

Go West, Young General,
Go West to the Colorado Rockies

To paraphrase someone, General Palmer, the founder of the 4UR as a guest attraction, was a different breed of cougar. Most young men, covered with earned and deserved glory and in good health after surviving such terrible battles, would look eagerly forward to a plush position with a large company in a large city where they could spend the rest of their lives at their private clubs recalling the cannons' roar or reciting their exploits to the eternal boredom of their grandchildren. Not Palmer. He had only begun his adventures. True, he had founded the Fifteenth Cavalry and was leader of half-a-dozen more units. True, history has badly underrated him, even though it has recorded all his accomplishments in the mighty war.

It is also true that at the age of twenty-nine Palmer craved other forms of action—and the way of that desire lay West, waiting for him. The far reaches of the western mountains held many feasts for the taking. All one had to be was ambitious, energetic and persistent. Young Palmer was all of that and more. For one thing, he was educated in many areas and many practical skills.

It is absolutely necessary to realize that this was a period of no income taxes, no environmental movements, no pension plans or endless withholding, no 401K retirement plans, no social security or affirmative action. It was all too big, too endowed with wealths of timber, precious metals and water beyond the eventual conception of waste. Everything was possible and the scope of the West seemed endless. There certainly was not a population of 260 million in the United States, nor automobiles past accurate counting. If a stream was saved from pollution in those days, it was because a powerful person thought it the right thing to do. If a town was planned for its future beauty and creativity, it was because of an extremely rare and powerful individual. Such a man was General William Jackson Palmer, the empire builder. At the close of the war, looking west with the eyes of a visionary, he instantly turned down the presidency of a railroad in the south. Instead, he accepted a post as treasurer of a portion of the Union Pacific Railroad, which was on its western trek to hook up with the Central Pacific, thereby linking the vast continent together.

His portion of the great endeavor began in Kansas City, with a line called the Kansas Pacific that was to meet the main line in western Nebraska.. In a letter to his uncle, Palmer explained:

> Young men without money can only make a fortune by connecting with capitalists. The heaviest of these reside in the East where they can look after their own affairs. But the best place to invest capital is in the West. Eastern capitalists must therefore have representatives here to attend to their interests if they wish to invest heavily in the West. Such representatives, if able and correct, must acquire great wealth and influence with their distant principles to a greater extent and more rapidly than if they lived in the East where the capitalist can judge for himself.

Palmer was honest, superbly talented and a proven leader of men, and was therefore handed a railroad, the plains of Kansas and a deadline. This would have made most men quiver like wind-blown aspen leaves. Palmer relished it.

That year, 1865, there were only 35,000 miles of rail in the entire United States. The figure would double over the next eight years. Palmer was in a furiously expanding industry. Nevertheless, Palmer spent a great deal of time in crucial office work. He made many trips to Washington to invigorate subsidies and secure grants for the Kansas Pacific. This is in direct contrast to what he would do in his later years for his own railroads, the Denver and Rio Grande. They were the only railroads built in this nation that used so little taxpayers' money.

Palmer had strongly urged the Kansas Pacific to look favorably on a southern railroad route across the country, to avoid the often extreme weather of the northern route, and he was chosen to map and survey the line. He was ecstatic. His party was made up mostly of former army officers from his own regiment. In fact, Palmer surrounded himself with these proven men of the Fifteenth Pennsylvania for the rest of his life.

In 1867, with Palmer in charge, and under a military guard, the party surveyed the plains in the summer heat. They moved on south to Raton Pass and down to Santa Fe, losing several horses to thievery on the way. In New Mexico, the surveying parties split, one taking the southern route through Fort Bowie (near present-day Willcox, Arizona) across the southern reaches of New Mexico and Arizona. Palmer led the other group west from Albuquerque along the thirty-fifth parallel, closely following what is today Interstate 40 and the Atchison Topeka and Santa Fe tracks.

In the rugged canyon country between Prescott and the base of the San Francisco Peaks (just around the mountain from present-day Flagstaff), they were attacked by Apaches. Palmer admitted they had foolishly been traveling in the bottom of a canyon. Though they had boulders for cover, arrows and a few rifle shots kept them pinned down. The general knew it was only a matter of time until they would be completely surrounded. He later wrote:

> How we got up, God knows: I only remember hearing a volley from below, shots fired above, Indian yells on all sides, the grating roar of the tumbling boulders as they fell, and the confused echoing of calls and shouts from the canyon. Exhausted, out of breath, and wet with perspiration, boots nearly torn off and hands cut and bleeding, I sat down on the summit and looked around. Everything was quiet as death; the Indians had disappeared—melting away as suddenly and mysteriously as they had at first appeared.

The Apaches were probably so surprised by the boldness of the uphill charge that they thought these people were inhuman. They had gone twenty-four hours without food, and were nearly frozen, in spite of clambering over twenty miles of boulders, before they connected with the rest of the party. The next day one of Palmer's favorite horses, Signor, fell off a cliff to his death. Maybe the Apaches won after all.

His other pet horse, Don, survived with him all the way to California. Even in a time of cheap horses, Palmer did not neglect his four-legged brethren anymore than the two-legged ones. He shipped Don by boat all the way back to New York, then west to the end of the tracks. Don, along with other animals, lived out his days at Palmer's Colorado home as a true and trusted friend of the former cavalry officer.

To delineate the general's character a little more, we'll return briefly to the Civil War. On a night of dark and dread, during the Civil War, a small black boy about thirteen years old appeared before the general almost like an apparition. He was a runaway slave. Standing as tall as his spine would allow, he begged Palmer not to send him away. Palmer told his supply officer to order a uniform and food for the boy. The smallest uniform still engulfed the frail frame and hung loosely, but he wore it proudly. The troops joked and chided Palmer about the "old man" who had joined up. From then until his death, the former slave was known as "Old George."

Palmer made Old George his orderly, teaching him those duties and also taught him to ride cavalry style. No matter what difficulties of terrain or terror of battle they encountered, Old George was by Palmer's side. He was there when they charged the canyon walls at the Apaches and also when Palmer first took a horseback tour of the majestic Pikes Peak area that he would build into a beautiful city. Old George was a confidante above all and was treated as such.

The party reached California in January 1868, having surveyed (along with the southern party) 4,464 miles of heretofore unmapped country. His horse, Don, floated easily back to New York on a ship, but Palmer rode horseback and by stagecoach to meet the end of the track for the Kansas Pacific. This endeavor alone was more than enough to make any man respected for a lifetime. Palmer was just beginning.

He tried with great enthusiasm to get the railroad directors to build a southerly route to California. It never happened, but (as occurs to most pioneers in any field) other railroads followed his survey stakes years later. However, because of Palmer's persistence the Kansas Pacific did extend its line to Albuquerque, El Paso and eventually Mexico City.

In those days, building a railroad consisted of almost endless variables and obstacles even in a time when many had an "I didn't see it" attitude. Money had to be raised from investors all over the United States and Europe. Congress was not innocent of corruption and had to be cajoled into allowing any one company into a city or over a pass. Of course, there was the enormous effort of physical construction as well. The huge Baldwin locomotives had to be shipped from the east coast and the rest of the rolling stock from all over Europe. The rails were made in Wales. The wooden ties were cut in mountains by railroad crews and had to be shipped a long distance when the end of the track was somewhere in Nebraska.

Above all this, the track crews had to be protected from resentful, raiding Indians and often-savage natural elements. In this age of specialization it is probably hard to believe that General Palmer had to oversee all of this.

And then he fell in love. Queen Mellen, a daughter of a prominent New York attorney, was "it" for Palmer. Naturally, they met on a train in the spring of 1869. He wrote to her constantly during their eighteen-month engagement. His love affair occurred during one of the most expansive and exciting times in the history of America. Miss Queen was both his love and his luck. Soon after their meeting he was elected to the board of Kansas Pacific, but the eastern-run company could never fulfill his cascading visions. At least a small picture of this man is visible from just one letter he wrote to his love. In January 1870, he wrote to Queen from aboard a train in Salina, Kansas:

> I had a dream last evening while sitting in the gloaming at the car window. I mean a wide-awake dream. Shall I tell it to you? I thought how fine it would be to have a little railroad a few hundred miles in length, all under one's own control with one's friends, to have no jealousies and contests and differing policies, but to be able to carry out unimpeded and harmoniously one's views in regard to what ought and ought not to be done. In this ideal railroad all my friends should be interested, the most fitting men should be chosen for the different positions, and all would work heartily and unitedly towards the common end. . . .

Then I would have every one of these, as well as every other employee on the Road, no matter now low his rank, interested in the stock and profits of the line—so that each and all should feel as if it were their own business and that they were adding to their store and growing more prosperous along with the Road. They should feel as if it were their own Road and not some stranger soulless corporation. How impossible would be speculation, waste, careless management on "Our Road."

It would be quite a little family, and everybody should be looked after to see that there was no distress among the workmen and their families—and schools should be put up for them, and bath-houses, and there should be libraries and lectures, and there would never be any strikes or hard feelings among the labourers toward the capitalists, for they would all be capitalists themselves in a small way, and be paid enough to enable them to save something, and those savings they should be furnished with opportunities of investing in and along the Road, so that all their interests should be the same as their employers'.

But my dream was not all of a new mode of making money, but of a model way of [life] conjoined that with usefulness of a large scale, solving with it a good many vexsed [sic] social problems.

A dream of Nirvana? A dream of Utopia? An impossible dream? To most it would seem so, but in five years' time Palmer had built just such a railroad, the Denver and Rio Grande, and an elegant city where in 1869 there had been only a few trappers and bands of hostile Indians.

As might be expected, historians still argue whether he dreamed of the city or the railroad first. In his letters to his beloved Queen, they seem to be the same dream. At any rate, the building of the railroad and the city certainly coincided.

Palmer's was the first railroad into the rugged, steep-sided southern Rocky Mountains of Colorado, eventually driving right through the heart of the 4UR realm and only a river bank away from the ranch's eastern boundary. It obviously made it far easier to haul rich ore from the mines to the smelters. The best example is this: When the train reached Leadville to the north in 1880, it replaced twelve thousand pack mules and all the attendants who slaved to keep them healthy and moving.

With some powerful friends, Palmer raised enough funds to construct a line about seventy-five miles long to Monument Creek at the base of Pikes Peak. The general had fallen incurably in love with the site at the foot of the famous peak. The little, lush valleys of Fountain Creek and Monument Creek were pictured in the visions of his dream city. The impossible became the possible. After all, Brigham Young had Salt Lake City built according to his wishes.

Palmer, with iron grit, was unwavering in his vision of a beautiful city, one where culture was supreme. At the time, it seemed like the obsession of a madman, as so many great accomplishments often do. There were only five thousand residents in Denver and many of them were gambling, mining and whorehouse rowdies. From the Mile High City to Santa Fe there were fewer than ten thousand *Homo sapiens*. Forget the bright or lucky few who had made it big—most of those people were scratching for a living as miners, lumberjacks or cowboys. These were not your usual opera-goers.

General Palmer enlisted the help of a Dr. Bell and Governor Hunt of Colorado and started buying land along Monument and Fountain Creeks in 1870 for $1.25 an acre. He planned every move, every street, every park, everything with great care and love. He envisioned two communities, Fountain Colony and Bijou Colony. They would later, of course, be called Colorado Springs. In just five years, Colorado Springs had twenty-five hundred residents, churches, schools, a hospital, an elegant park and even a college. Trees had been properly spaced along the city streets, which were designated to be at least one hundred feet wide. There was an extensive gravity irrigation system serving every lot in the city.

His wife, Queen, had founded one school all on her own, as well as their famous sixty-room mansion in the hidden valley behind the majestic rocks of the Garden of the Gods. The first move the general made was to hire a forester to prevent settlers from cutting trees in the Garden of the Gods.

He enlisted a Scottish landscape expert, Mr. Blair, to survey the valley to plan the "Castle" and tailor the grounds. Old George accompanied them on this most special of countless special occasions. As they rode, a great eagle winged upward from its nest high on the

Glen Eyrie, Palmer's castle in Manitou Springs. (Courtesy, Western History Department, Penrose Public Library, Colorado Springs, Colorado)

bluff. It spiraled above them awhile and then plummeted down, finally breaking its descent with its mighty wing feathers, so close they could see a yellow eye glistening in the sunlit air.

"I don't think he wants us here," Old George stated emphatically. Mr. Blair thought otherwise. He watched the eagle's circling ascent and then glanced at Palmer. They both had great vision and courage.

The decision had been made by the eagle, in truth, even though it was Mr. Blair who said, "This valley should be called Glen Eyrie. In Scottish that means Valley of the Eagle's Nest."

"And so it shall be," said the general.

Once the decision was made it is not hard to imagine Palmer's heart pounding with love and joy, just as Charles H. Leavell's heart would do much later at his first sight of the 4UR. In February 1872, the home on the glen was finished, overlooking the creek that ran through Queen's Canyon (which he had named after his wife). Palmer gave names to all the rock prominences. There was wild game and all manner of exquisite natural things to observe in awe and dignity. The castle's setting was as splendid as any in the world. From its tower room one looked out over a paradise of unspoiled beauty.

Late in 1872, a new child's voice was heard at Glen Eyrie. The general's first child, Elsie, was born. He loved all children and he and Old George took them on endless expeditions and nature studies throughout the Eagle's domain. In 1880, a second little daughter, Dorothy, arrived to join the happiness in the valley of the eagle's own survey. Just five years prior to her birth, Palmer had acquired the 4UR. It became a satellite residence where he entertained notables from all over the world, just as he and Queen did at Glen Eyrie.

Colorado Springs was almost entirely designed to Palmer's personal specifications; anyone who wanted to live there lived by his rules and built to his visions. He owned all the land to begin with, tens of thousands of acres. He donated the land for the churches, hospital, schools and parks. No road or trail was put in the parks without his personal approval. The man who single-mindedly opened millions of acres of wilderness to railroad traffic showed a different side in Colorado Springs.

The last spike of the Kansas Pacific Railroad was driven in August 1870, when crews working east from Denver met crews working west from Kit Carson. The four hundred men of Palmer's crew had worked themselves nearly to exhaustion on the line, and when it was finished, a train brought the men a fine supper with some fresh California grapes and peaches, ice cream and other luxuries, and even cigars to close with, but no whiskey. The strait-laced Quaker in General Palmer vehemently forbade alcohol to enter the celebration. Anyone who has ever been around mining and railroad camps knows that he probably wasn't totally successful in this denial, no matter how much his men respected him and his religion.

Though the Kansas Pacific was content to let the rest of the West lie fallow, Palmer wasn't. Here was the opportunity he had been waiting for. Here was the chance, in the fall of 1870, to create the ideal rail line he had dreamed of, on a train, so long ago.

To put things in some kind of perspective, it should be remembered that monumental events came and went quickly. The West—after the Civil War—didn't evolve, it exploded like a volcano of vast opportunities. Palmer at this time was one busy general, but again he was only directing a great orchestra in rehearsal. The main performances were upcoming. Colorado was holding its arms out, filled with rich gifts for those who knew how and when to accept them—and Palmer knew it. His visions were clear almost every time.

Although Palmer saw without question the growth that was to come to the area, not everyone did. It takes money to build a railroad, and it had to be raised. Palmer promoted some of the capital on his European honeymoon—a classic mixture of business and pleasure. He knew that it also takes friends, earned friends, to succeed. His father-in-law secured more money in New York, and his friend, D. Bell, raised some too.

The Denver and Rio Grande line reached Palmer's dream city of Colorado Springs on October 21, 1871. The inauguration was ceremonious and probably a bit pompous, with speeches from the governor and other dignitaries. Palmer wasn't there. He had more important things to do. He was riding a mule in northern New Mexico's Valle Vidal searching for a pass through the Sangre de Cristo Mountains. Dollar by dollar and mile by mile, the line continued. It reached Pueblo, Colorado, in June 1872.

The dreams continued. Palmer had envisioned a railroad traversing the length of Mexico and he intended to see it done. So, as his own railroad crept against mountains, with a certain amount of red ink staining his ledgers, he and Queen traveled to San Francisco and sailed to Mexico to sound out the government on the project. This first trip left him frustrated, but produced a key contact, one of upcoming great power. Palmer met a man on the ship to Mexico—a quiet man, very polite but mostly withdrawn. He and Palmer shared some conversations and a few acknowledging smiles but little else. When the ship docked in Manzanillo, Mexico, the man and his companions vanished. It wasn't until years later that Palmer discovered this shipboard companion had been Porfirio Diaz, who was, at that time, a rebel with a price on his head. Later, of course, Diaz was to become president, then dictator, of Mexico for thirty-five years. Fortuitously for both of them, each remembered the journey they had shared. Years later this fateful shipboard meeting led to Palmer's initial founding of the Mexican National Railway.

Back in Colorado Springs by the summer of 1872, a consortium led by Palmer bought the 40,000-acre Nolan Grant on the south side of Pueblo. He then founded South Pueblo, with its model community for its workers, and the Colorado Coal and Iron Company. (The Colorado Coal and Iron Company eventually became today's Colorado Fuel and Iron Company, more commonly known as CF & I.) In 1880, the South Pueblo Ironworks was founded and by 1882 it was producing rails for western trains, where before they had had to be imported from the east at nearly double the cost.

For several years, Palmer's life bounced between trips—mostly fruitless—to Mexico City, development of both the Denver and Rio Grande and the city of Colorado Springs and working vacations on his 4UR Guest Ranch. By all rights, the Denver and Rio Grande's plan to push over Raton Pass into New Mexico and establish the long-sought warm route to California should have been a cinch. The tracks were already within a few miles of Trinidad, just north of Raton Pass. But the money sources dried up, and Palmer found himself with a nice little railroad going from near Trinidad to Denver but far from the continent-busting enterprise he had envisioned.

Then the Atchison, Topeka and Santa Fe Railroad (as it was later called) changed hands. New money saw an opportunity, pushed track into Pueblo and prepared to send more rails, not only over Raton Pass but up the Arkansas River as well. As sometimes happened in the West, the operators of silver mines at what later became Leadville discovered one day that they had been throwing millions of dollars' worth of lead into the mine tailings. A bonanza awaited the rail line that could get there first.

What happened next was close to disaster. The crews pushing the Denver and Rio Grande up the canyon built a fort of logs and rocks, armed themselves with rifles and then held back the laborers of the Santa Fe. Both camps were heavily armed and edgy, but miraculously no shots were fired. This standoff continued for many months while

Palmer went to Boston to try to head off a bloodbath. Then the battle shifted to court. After months of one decision being overturned by the next decision, the courts decided that the Santa Fe had the right to cross Raton Pass into New Mexico—but it had to give up the route to Leadville, which was left to the Denver and Rio Grande.

So there was Palmer, with his way to the south blocked and his nose pointed west into the big mountains. This wasn't exactly what he had planned, but it seemed to be the way things would have to be. No one had ever built a railroad into mountains as steep and dangerous or as high as these Rockies, not even with the narrow gauge.

Palmer loved the three-foot narrow-gauge track. Not only did it give the train a lower center of gravity, but it also allowed the train to turn more tightly. Narrow-gauge track and stock were 37 percent cheaper to buy and operate than standard gauge, for which the track was four feet eight inches wide.

The legal way into the mountains was cleared. In his private car in Pueblo one day, the general called in his top engineers for a conference. It seemed that new mines, back in the Rockies, were being developed each afternoon, and Palmer wanted to haul the ore out of those mines. These men—nearly all of them close friends from the Civil War—spent hours patiently explaining to Palmer why it would be physically impossible to build a railroad to Leadville. All the while, Palmer sat quietly. When they appeared to have exhausted their reasons why this couldn't take place, Palmer stood and said, "Gentlemen, the decision has already been made. Do it."

About that same time, Palmer and a few others were granted the concession to create the Mexican National Railway. One track would head west from Mexico City to Manzanillo on the coast, while another—the big one—would start from that track west of Mexico City and go north to the Rio Grande at Laredo, Texas. The track to the coast was completed. Then the north-south railway crews started from both ends, but they were not going to meet in the middle. With a three-hundred-mile stretch of country still separating the ends of the track, Palmer found he couldn't finance the rest of it. The Mexican government took the line over and finally completed it in 1886, but Palmer is still honored and acknowledged today as the founder of Mexico's national railway.

Back in Colorado, the Denver and Rio Grande found itself sending spider webs of track all through the Rocky Mountains. Every time some prospector discovered rich ore, the miners moved in, and the spur track to the mine wasn't far behind.

In 1878 there were 373 miles of track on the Denver and Rio Grande. Four years later, there were 1,281 miles. In 1881, Palmer founded Durango, Colorado. The smelters were built there because of his railroad. He pushed track from it through the tortuous Animas Canyon to the mines at Silverton the following year. His railroad extended as far south as Espanola, New Mexico, on the Rio Grande River.

For as long as General Palmer ran the Denver and Rio Grande, by the way, the Southern Ute and Jicarilla Apache Indians (through whose land some of the track went) rode the train free, at his orders.

In 1883, Palmer faced hostile stockholders in the railroad. They voted some of his army buddies off the board of directors and had the votes to oust him as president of the line. In anger, he beat them to the punch and resigned. Two years after he left, the employees of the Denver and Rio Grande called a paralyzing strike that sent the railroad into receivership. It changed hands several times in the next eighteen years.

At the time of his resignation, General William Jackson Palmer was forty-seven years old, the builder of a beautiful city, at least three towns, two steel mills and three railroads. Palmer had effectively, and relatively benignly, opened vast mountain wildernesses to commerce and to elegant excursion trains full of sightseers from the East. He was living, with a loving family, in a cultured environment at the base of his beloved mountains, surrounded by lifelong friends who would gladly take a bullet for him. He was worth much more money than he would ever need in his lifetime. Life was quiet and serene and good. So he started another railroad.

The Western terminus of the Denver and Rio Grande was Grand Junction, Colorado. The year he resigned from the leadership of the D&RG, Palmer founded the Rio Grande Western; by the late 1880s, he had pushed it from Grand Junction through to Salt Lake City. It would be fruitless, in this account of the 4UR realm, to cite any more of his enormous accomplishments, except, of course, in their proper place and time—his building of the railroad to Creede and the founding of the 4UR Guest Ranch.

Much earlier, Onate, de Anza, Zebulon Pike, Kit Carson, Fremont and others had, at great sacrifice, opened minds and trails that would make Palmer's mighty works possible. Those works would wind and turn and finally help other latter-day pioneers such as the Phippses and the Leavells to find their own courageous way throughout the Rockies and the world—with all of it pointing at the 4UR.

CHAPTER 13

Charles H. Leavell's Power on the Ground and in the Air

Charles Leavell's company built the $15 million Rancho Seco Nuclear Power Generating Station for the Sacramento, California, Municipal Utility District. It also constructed a $5.5 million industrial sewage system there. The Leavell Company was soon awarded more classified building contracts for the Atomic Energy Corporation, such as five classified projects for the AEC at Jackson Flats near Las Vegas, Nevada, as well as huge projects ranging from nuclear reactor maintenance to vast water treatment centers in Connecticut and Omaha, Nebraska. It would be almost impossible to list all the projects touched by Charles H. Leavell's influential expertise and his company's expansion during these middle years. However, a few were extraordinary and should be related.

The aforementioned Rancho Seco Nuclear Power Plant at Sacramento was special in the sense that it took all of Leavell's people's engineering experience and expertise, accumulated over the hundreds of past projects, to pull it off. The CHL Company was the sponsoring contractor on this joint venture with Dravo Corporation of Pittsburgh, Pennsylvania. They pulled together binding, intricate specifications for the design, concrete, steel, and earthwork. The critical project came in ahead of schedule, exactly on budget and, amazingly, with a flawless safety record. Again, Charles gives most of the credit to others.

In early 1963, Charles had contracted design work for a vast extension of Bangkok, the principal city of Thailand. It was to be across the Chao Phya River, the major waterway flowing through the city. They had finished a lot of very astute and long-range planning for the Thais, which later became the basis for the huge growth of the city. An exciting success indeed!

Shirley and Charles were making their final, exultant, hard-earned, celebratory trip back to Thailand. They had been informed that the Thai king, Blumibol Adulyadej, was going to name Charles as the Honorary Consul General to Texas from Thailand and that he would he presented at court. Of course, he and Shirley knew they had certainly earned this gesture of appreciation, but even so they were excitedly looking forward to it.

The El Paso couple were met at the Bangkok airport by a distinguished retinue, with bounteous and beautiful flowers for Shirley and several limousines full of official greeters. With ceremony that would have pleased the king of any land, they were escorted to the only hotel of any consequence in town at that time, the Arawan. Of course, with a large initial boost in planning from the Leavells, there are now many skyscrapers and hotels all over.

Their assigned host, Osot Kosin, Minister of Foreign Affairs, asked Charles if he had proper attire to be presented to the king the next evening. Charles was panicked. Shirley, in contrast to what most would expect, appeared as cool as all the flowers that decorated their suite like a lush garden. She was right. Mr. Osot promptly brought to their suite the royal tailor and the royal seamstress. They were expertly, almost effortlessly, measured for evening clothes. No, they were more than that; Charles was precisely fit with complete tails and Shirley with a beautiful Thai dress of the finest silk. By five o'clock the next afternoon, prior to the presentation, Charles recalls, "We were in the most gorgeous setting and were being taken promptly at eight o'clock to be presented the king. It was a very formal affair and turned out," as Charles understates, "most interesting."

The king bestowed upon him the recognition of being the Thai Honorary Consul General to Texas as well as El Paso, and as a surprise bonus awarded him the honor of the Thai White Elephant, third class— very close to the top indeed. (He is now first class.) Charles felt like he had been awarded an Olympic bronze for the decathlon. Their

Charles Leavell becomes the Thai Honorary Consul General to Texas.

daughter, Mary Lee, has now succeeded Charles in this post. Charles remembers clearly how proud he was that evening to have both his and Shirley's priceless work for the Thais recognized so graciously.

Everything was going smoothly until just before he was presented to the king, when Charles had a moment that could have caused the lovely event to be a disaster. He was standing next to a Thai gentleman who was dressed in a cloth that looked to Charles as if it would do for the cover of a pool table—very heavy green cloth. Not dreaming that the man understood English, Charles quietly expressed his thoughts,

"Boy, you're really tough to wear those heavy clothes in this warm and humid weather."

The man replied, in perfect English, "Well, I feel like a jungle monkey in this suit and this is the first time I've worn it since I left Texas A & M." It turned out that he was a Texas A & M graduate and was wearing the Texas school uniform, despite his great discomfort, in Charles's honor. He also was the king's brother.

On the Leavells' second visit to Bangkok, the royals once again held a banquet to honor them. It was a beautiful and overly bountiful banquet in a decorative, Oriental-paneled dining room filled with servants. To the Leavells' surprise, the servants dispensed the feast on their knees with all the royal courtesies.

The host again was Osot Kosin, Minister of Foreign Affairs, who spoke in his best Texas A & M English and then repeated himself in Thai as a reverse interpreter. He stood and proclaimed, "Our toasts in this country are always about the life of the people we are honoring. And I would like to talk about Charles Leavell." Which he did, in a very complimentary and soft-spoken manner. At the end of the toast, he said, "Now ladies, from now on this is a men's dinner; your chauffeurs are waiting for you and your attendants will take you back to your hotels and your homes. Now please gentlemen, let us rise and bid good evening to our ladies."

So Shirley reluctantly rose and left with the other women in a mass exit. The big sliding paneled doors had hardly closed behind them when the most beautiful Japanese girl one could possibly dream up sat on Charles's lap. Then a Thai girl of equal beauty and glow joined her. Their soft, delicate flesh used as their chairs one knee apiece of the astonished El Paso contractor.

An orchestra entered and started playing and the other men began to dance with their girls. Charles did not dance with his. He, like any other feeling man, had confused thoughts and kept glancing at the paneled doors, hoping the most beautiful girl in his world, Shirley, would reenter and legitimately save his soul. Even so, his thoughts cleared enough to tell himself he had damn well better head for the hotel. He tried to be polite and stay long enough not to offend his most generous hosts. Then he excused himself and went back to his hotel, drenched in a sweat that had nothing to do with the humidity.

He says, smiling at the situation, "I remember Shirley was sitting straight up in bed in our Arawan Hotel suite and she said without wasting a second of time, 'Charles, tell me exactly what happened.' Well, I didn't have much to tell her." And he wisely ended his explanation right there, with his usual right-to-the-point brevity.

This special time led to many return pleasure trips for Charles and Shirley. They made several great tours of this now burgeoning Pacific Rim powerhouse and also made many lifetime friends there. Charles says with much feeling, "And I am still very, very fond of them. They are very dear to me."

In 1968, the Charles H. Leavell Company was awarded the contract to build a fourteen-structure complex of professional buildings in Liberia, Africa. Liberia had once been an American colony where freed slaves, particularly from the southern United States, were expatriated and given farmland, in hopes of helping themselves and the new country. Most of the blacks who emigrated there had American names and all spoke pretty fair English.

Charles was on his way to visit the job sites, which consisted primarily of educational buildings, including a college campus, a big stadium and other accessory buildings, for the first time. He and Shirley had boarded a 707 and flown a long trip through Europe and back down the coast that led to Liberia. They landed at the airport, which was about seventy-five miles away from Monrovia, the capital. It was an old World War II military airport that had been converted to commercial use by the Liberians. There was a good highway back to the big city of Monrovia.

When they entered the immigration department, there was a black man sitting on a high stool. Charles relates, "I mean he was *something* and he was very quiet and very severe with me, saying, 'Sir, you do not have a visa.' I said to him that I was told in Washington that I did not need a visa. He said, 'Well, I'm in charge here and you must have a visa, and so does your wife or you are not allowed in *this* country.' Then I asked him, 'What am I to do?' and he said, 'Well, either get back on that plane or wait here in this office until the next plane comes through.' I was getting pretty anxious and I asked him when would the next plane be in? He said, very abruptly, 'In four days.' Then I said to him that we can't get back on that plane because it's going to Ethiopia and from there to India. There is no way we can do that."

So he and Shirley were incarcerated in a little room about six feet by ten feet, with one straight-backed wooden chair and one iron cot with dingy sheets wrapped around a straw mattress. Charles was finally allowed to use a telephone and he called his office, which was seventy-five miles away. The manager, who had been unavoidably delayed himself, was on the road and arrived two or three minutes later, intending to pick them up. After hearing their predicament he said, "Mr. Leavell, I will get you a private plane and get you folks straight out of here. Now don't worry a minute more about it."

When the manager finally called back from Monrovia, he said, "The best I can do is get the Minister of Foreign Affairs to give me a letter authorizing your entry into Liberia. I'll be there at 6:30 in the morning with a private plane."

Then all Charles and Shirley could do was sit in the room where they were locked up. After a bit of time spent encouraging one another, Charles gallantly turned back the sheets, preparing their bed with a grand gesture. Several large and small bugs were having a corporate meeting on the mattress. Shirley refused to lie down with them and sat up in the straight-backed chair all night. Charles had slept on bugs before, on his many back-country excursions, so he brushed those in sight toward the wall, pulled the sheets back down, and reclined.

With little real sleep, the Leavells were awake and waiting when, after seemingly many nights' worth of hours, the company pilot flew in with a very nice two-engine propeller plane. He also held the promised letter from the Minister of Foreign Affairs that said, "Please admit Mr. and Mrs. Charles H. Leavell for a period of seven days into our country and give them proper visas." Charles took the letter to the high-stool man and he stamped their passports. With great relief they got into their plane and were soon visiting the construction projects.

As it turned out, things were moving along on schedule and budget. After only four days the Leavells were ready to leave—three days ahead of their allotted time. However, they had to catch a certain plane because it was the only one to New York for the week. When the two weary world travelers got back to the airport, looking forward to a hasty exit, they were a little shocked to see the same arrogant official sitting on the same stool and, as they immediately found out, ready to utter the same words as before: "You have no visa."

Charles said, having great difficulty controlling his voice, "I know, sir, I have been through this with you before. Here is the letter from the Minister of Foreign Affairs."

The high-stool man looked at the letter and said, "This does not exit you from the country until the seven days are up. It has only been four days. You can't go."

Charles said, "You have to be kidding me." He was getting angrier and angrier by each frustrating moment. Of course, he knew that a fast exchange of hundred-dollar bills would have solved their problems from the beginning and made everyone polite and helpful, but because of the man's arrogance, Charles just could not bring himself to hand over a bribe. Ah, the price of honor!

Charles didn't know what to do. Their luggage had already been loaded onto the Pan American plane. He asked the man politely once more if he and Shirley could simply walk out and leave the country forever.

"No."

"What are we to do then—"

With even more arrogance than the El Paso couple had thought possible, the high-stooler said, "That's your problem. You stay here for a total of seven days. Simple."

Charles does not quite remember exactly how he blew up or just exactly what he said to the man, but he must have made the right threats, for suddenly the officious clerk scrawled in Arabic across his passport, "Exit without my personal permission and absolutely violating all the laws of our great land." Charles has kept that scribbled-on passport all these years. He still has his honor in place, but if pressed, could he truly say it was worth it? What he did say was, "So much for Liberia."

The story of the Khartoum adventure in worldly building is one of tragic irony. Of course, it didn't start out that way. The country of Sudan had secured financing from the AID program and awarded the Leavell Company a contract for a large industrial infrastructure at

Omduman, which is the old native city across the Blue Nile from Khartoum—the city so many wars have been fought in and over throughout recorded history. Charles was soon to have his own personal war tied to a much larger and tragic one.

At the time, Omduman was occupied entirely by army people and Arabic blacks. There were only seven Arabic states and because Sudan was the lowest on the list, they had many supply problems, lots of red tape and bungling that was plain natural to the land at the time. There were also transportation difficulties beyond imagining.

The Leavell Company had just completed purchasing all the material for the job. It was being shipped in two Greek vessels to a Sudanese seaport five hundred miles from Khartoum, where it was to be moved by rail to the building site. This project required a lot of complex machinery, huge pumps, paving equipment, steel materials of many kinds and other big components of all varieties and materials. Most had been purchased in the United States, but some was coming from Europe as well. All this massive, fearfully expensive material was on the high seas, headed for Sudan, when the Six-Day War broke out between Israel and Egypt. At this same time, in 1967, a revolution progressed in Sudan itself, headed by the famous revolutionary Numeri. Charles said this black Arab was on a par with Pancho Villa—certainly indirect, but a compliment just the same. Unfortunately, Numeri killed any and all who gave him even the slightest trouble.

"Everything around us was going to hell in a busted basket," Charles states with still very clear hindsight. "First, one of our supply ships was sunk in the Suez Canal and another was stopped by the Greek owners and taken back to Perius in Greece."

When Charles frantically communicated with the ship's owner, he was told, "Read the small print, buddy." Charles did and found it to say, in effect, that in the event of war, both the cargo and the ship belonged to the ship owner and he was free to do anything he wanted with them. So the Leavell materials, costing fortunes, were either lost or virtually so.

Then their communication lines with the job site broke down. All American and Europeans had left the country. Charles thinks there were only about twenty-eight white people left in all of Khartoum and six or eight of those were from his management team. For a month they tried desperately to wire, telephone, send paid messengers and everything else they could think of to reach his people. Charles could not get a passport into the country because of the wars. The U.S. State Department stopped payment on the job because, like the Greek ship owner, they said that in the event of war they had no responsibility.

Charles never quit. It just isn't in his genes. He found a very obscure German airline unfathomably owned by the Israelis. It was extremely dangerous going in, so Shirley insisted that he not go by himself. She was so adamant about it that Charles didn't dare buck her intuition. He took Bill Abrahamson, a big, tough, very fine construction man. In fact, Bill seemed eager to make the perilous trip. They finally got as far as Egypt, where they were unceremoniously incarcerated and placed in the Hilton Hotel with all windows boarded up and doors secured under heavy locks. They were held there for five days and then, with no explanation, were freed to get on an exotic flight that Charles had found out about. At last they landed safely—at least for the moment—in Khartoum. They had no visas, no anything really except their passports and their most polite attitudes. Happily, they were treated courteously in return. The country was still in revolt, but Charles talked his way into a meeting with the Minister of the Interior and his associates. To Charles's amazement, the minister was only about twenty-two or twenty-three years old.

Charles recalls, "He was one of those fly boys who piloted airplanes in and out of Sudan. And I remember very clearly that the only thing he had on his desk was an army automatic pistol which I occasionally eyed while I talked business with the 'pup' of a minister. It was a rough go from any angle you care to look at it."

The Leavell Company had no project money left, having spent millions for the confiscated shipboard materials. There was a very real danger of its going broke on this single project. The possibility seemed totally unacceptable, after having succeeded at almost uncountable sites and often extremely difficult projects of every kind in locations around the world. So Charles began to talk with all the powers of persuasion he possessed. He asked the Arabs, in the most persuasive of voices and reasoning, to use their own money to refinance this broken situation and brazenly, but sincerely, asked for payment to be made in New York in U.S. dollars at a U.S. bank of his selection. It was a bold move. His confidence had been bolstered by the fact that Shirley had insisted on joining him. Hence he had her reassuring, private support and much-respected wisdom, as well as the staunch, unfailing loyalty of Bill Abrahamson. Nevertheless, he faced the Arabs alone as he made his demands on this deal, which had most of the blood of his company flowing through very thin arteries—to say the least.

Charles was outnumbered one dozen to one by these handsome men in beautiful flowing robes as they all sat around a great, elaborately carved table. He waited several breath-killing, critical minutes as they softly discussed, back and forth, his life and welfare in a language of which he understood only a few words. At last, through an interpreter, they said, "All right, Mr. Leavell, these are the terms. Here is the deal. The money will be deposited in the bank you have named, payments will be made as you request, and the new time schedule will be set up. We are restructuring the contract that will have nothing to do with Washington or AID money at all. It is our money and we are going to pay you."

Charles took the deepest breath of relief in his entire life and then was jolted again as one important Arab added, "There is only one stipulation, the one requirement which you have to meet is: you must agree that you will personally stay here and supervise the job to its completion."

Charles said, with what must have been great trepidation and even pain, "Gentlemen, I cannot do that. Not only do I have to go run my other businesses, but I'll be needed elsewhere around the globe to assure the safe and timely arrival of our materials. We are talking about at least two years for this job to be properly finished."

They said emphatically, "Those are the terms."

So now it was Charles's turn to make them wait. He thought and he thought, doodling with his pen with figures on a paper to give himself time to figure a way out. Finally he agreed.

The next night, at two o'clock in the morning, with the steady Abrahamson by his side, Charles took Shirley to the iron gates, past many men armed with submachine guns, and gave her into the hands of his remaining people. The iron gates slammed shut between them with the clanging sound of doom. He stared numbed through them, watching and knowing that his beloved Shirley would soon be safely out of the country but that he was still there for no telling how long. It turned out to be about ten days. He talked his way into a temporary release and joined Shirley in Germany, where she was staying with her sister and brother-in-law, James K. Polk, a four-star general in charge of the entire U.S. Army in Europe.

After rapid consultations with various powers around the world, and the banks in New York, Charles learned that the money to finish the tragedy-plagued project had been safely deposited. Then the Leavells flew home from Frankfurt. The job was properly organized and all difficulties seemed trivial from then on.

The financial stability, and therefore the power, of the Leavell Company and of Charles himself, was saved. It had been a close, tough call. The result gave the city of Khartoum a modern industrial sewage system encompassing over four miles of full-force main line and an eighteen-mile collection system, including two lifts, two pump stations and a treatment plant that no doubt has saved countless lives from disease over the years.

After all the war and revolutionary obstacles, the Leavell Company astonishingly made a little money during that same year of 1970. That was also the year Charles finally got his real estate proposal from Allan Phipps to write a huge check to open the way to the dreamed-of paradise escape in Wagon Wheel Gap, Colorado. The 4UR always laid torturous trails for those who finally found succor and replenishment there.

In 1971, the Cold War and the hot Vietnam conflict were all at their peak. Early diplomatic negotiations were under way between the United States and China. Of course, Mao Tse-Tung, the communist leader, had died, and his successor government, the Gang of Four, had just been ousted as rulers of China. Charles's alma mater, Stanford University, had just concluded a treaty with the Chinese to send over a small delegation of graduates, with an intellectual background on China, for the purpose of studying Chinese education in general and higher academic standards in particular. Then the special group was to write a white paper of corrective recommendations. The group was headed by Dr. John Lewis, Dean of Political Science at Stanford and a world-renowned scientist. Charles and Shirley Leavell were selected to be part of the group of ten because of their prior visits to and intellectual interest in the vast, complex country.

On arrival they were joined by six communist leaders of cadres, all in severely simple uniforms, highly disciplined and especially polite to the select American group, but all suspicious of each other (a trait common to all communist regimes). China is, to make an understatement, a large country of more than a billion people. To Charles it seemed like a village enclosed by ancient walls of the spirit.

Charles says, "China has charm—the greatest being its people, a lively, humorous, intelligent race. The cities are quiet and dark at night, but they are still full of walking and bicycling people. Thousands and thousands of them all with a mission of their own. At that time there were fewer than one thousand lawyers in all of the huge land called China—none of them criminal lawyers. Peer and family pressure keep the Chinese orderly, but severing a hand for a minor theft or jailing a person for a slight political dissent has a certain deterrent value as well."

Charles had told a close friend in the group, Bill Scott, that he would like to go out very early some morning and feel the lifeblood of the country flowing before the full restrictions of the day came to full force. They got up at 3:30 in the morning, taking a map of the city of Beijing and the address of their Russian-built hotel written in Chinese as a precaution. They walked the silent, but already crowded, dark streets until dawn, observing the Chinese queue up for their daily share of food, fresh milk and wood. There was an eerie, ghostly, even ghastly feeling from so many people and so little talking. There was no singing, no whistling, no sounds but the haunting shuffling of thousands of sandal-clad feet and the bicycle tires on the gravel streets. It was as if they had hidden their souls even from themselves to prevent further violation. Leavell and Scott visited a Buddhist temple and watched the women pray voicelessly. They saw old men doing Tai Chi exercises as precise and disciplined as the most specialized military force drills. Charles and his friend had felt these people all right, out of their nature—out of their human nature—but they knew a great spirit of intellect and dedication was hidden under their fearful, obedient surfaces.

As the sun rose above the city, large military trucks full of soldiers and arms shattered the silence like a dropped glass pitcher. The personnel carriers were full of Chinese veterans of the Vietnam War. The people felt free to show and sound their voices at this. They ran to the trucks and cheered them home. The before-dawn walk of two old friends had revealed more about the country than ten thousand so-called statesmen.

The Leavells' travels abroad were not all business and terrible deadlines of construction and completion. They toured China and the Inner Kingdom of the Great Wall—and in fact, most of the available world—for knowledge and just for the plain fun of it. Sometimes, though, even the latter turned into greater adventures than even they could have expected.

Of one safari among many, back in 1961, Charles speaks fondly, "I was young and vigorous and an avid big-game hunter." He and his son, Pete, both got the big five (lion, elephant, rhino, leopard and cape buffalo) on his first safari, for which he had enlisted the famous hunter, Frank Miller. It had been an adventurous time there in the bush, with all the wildness and excitement it generated, but the climax came in the third week when they were pursuing a trophy rhino. They had spotted him many times, but now saw him close in a field of *siskel* (large cactus plants) listening to them. They were walking forward into the animal's personal territory—hunter/gun-bearer and Charles—when suddenly the two-ton monster charged from the cactus bush. Charles took dead aim at its spine above the lowered head and pulled the trigger on the 405. Even before he raised his gun, he remembers vividly his white hunter's pants quivering. The gun double-crossed him and misfired. The gun-bearer ran to the side and diverted the animal's charge; a lucky thing, or else the many pieces of Mr. Charles H. Leavell would still be floating somewhere above Africa. The gun-bearer ran to the Land Rover and grabbed Charles's other gun. Charles's bad leg prevented him from doing so himself, and he was left very vulnerable. They were now surrounded by six territorial rhinos. The bearer handed Charles a

Stanford Alumni Association visit to the People's Republic of China: Betty and Bill Scott (back row), Shirley and Charles Leavell (left front), Gladys and Joe Moore (right front), 1978.

460 Magnum. He shot and killed the monster and the others scattered. He then knelt down and cried for both himself and the bold rhino.

On another occasion they were stalking buffalo in very heavy brush. They were all crouched over, hunters, bearers and all, moving single-file along a dim game trail at jogging pace. Shirley was in the rear of the line. She had been thoroughly instructed against speaking or coughing or making any human noise. She understood the instructions, whether the great white hunters believed so or not.

When Shirley heard brush snapping behind her, she turned and saw that a big-horned rhino had gotten in line behind her. She didn't yell. She didn't cough. And she sure as hell didn't run. She faced the rhino while tapping a tin cup on her knee as hard as she could, trying to get the other hunters' attention. They turned to her. Before anyone could shoot and rescue the "damsel in distress," the rhino suddenly whirled and bolted away as if in great fear of the noise from the tin cup. It's almost certain he had never heard that noise before and very possibly never would again. Yes sir, Charles's Shirley is a beautiful woman . . . and a very brave one as well.

None of Charles's family had ever heard of a "Sing Sing"— except, of course, the infamous U.S. prison—until they safaried to the wildest place of all, at that time: New Guinea, home of headhunters and jungle insects bigger than a grownup's hand. In 1978, that's where Charles took his family—Shirley, Pete, Mary Lee, and her son, Fred, who was and is a powerful young man despite paraplegia caused by a broken back from a ranch accident.

They had engaged, as an outfitter and guide, an Australian, Keith Buxton, who had fought the Japanese during World War II in Papua, New Guinea. They took both an air and a land safari in that wild aboriginal country. They witnessed and lived through several "Sing Sings," which turned out to be frenzied dances done by native warriors all decked out in battle costumes, feathered and beaded headdresses, and all loaded down with spears, bows and arrows and their cheeks stuffed with the betel juice "narcotic."

Later the Leavells were driving in a minivan through the Highlands of New Guinea when suddenly before them was a burning village and a tribal war in progress. Buxton, who was known there and could speak in dialect, bravely advanced on foot. All expected him to be beheaded first and then the Leavell family next. To their surprise and great relief, Buxton soon returned and informed them that the battle would halt until they had driven through. As they passed the smoldering village, a tribal chieftain ran after them. They stopped the van. He stopped and exuberantly doffed his headdress, stuck his head into the vehicle, as all held their collective breaths, and said, "Mr. Buxton, has my membership in the country club been approved—"

The entire Leavell family ached for Buxton to scream "yes." To this day they don't know what he said in dialect, but they assume it was affirmative.

Soon afterward they were watching another wild and interminable "Sing Sing" in the Kofue village of Tufic. There were no roads at all there and all travel and transportation, in and out, was by hand-hewn outrigger canoe. These remote people lived at that time exactly as their forefathers had for thousands of years. This "Sing Sing" lasted all night, much to the Leavell family's weariness.

The next morning Shirley was showering in their house on stilts, using their old G.I. canvas shower bucket. It ran out of water about the time she was well-soaped. Charles shouted for some female member of the family to go help out with a bucket of water. Soon an old native man came trotting up with a bucket of water and a ladder. He promptly barged into Shirley's bathroom. Before she could recover from her astonishment, he set up the ladder and proceeded to climb it and douse her nude body with the bucket of water. As always, remaining cool under fire, she shouted to the old man loud enough for all the New Guinea villagers to hear, "I am clean!"

All the Leavells mentioned here have continued to do their own kind of "sing sings" ever since.

CHAPTER 14

The Leavell Company Joins the Cold War with Hot Projects

Some of the dams the Leavell Company built may later have adversely affected the environment (who knows—), but the great waste and water treatment plants, and the efficient, often beautiful health and educational construction projects most certainly contributed to the health and welfare of countless thousands of people around the world. The enormity and number of these life-giving projects are staggering. Charles Leavell's company even built a third dome for the McDonald laboratory at Fort Davis in 1968 for the University of Texas, housing a 107-inch telescope to study the solar system prior to the first U.S. landing on Mars.

When the John F. Kennedy Medical Center, elementary and high schools were completed in Liberia in 1968, one of the project managers summed it up precisely in a letter to Bill Abrahamson (who was vice president in charge of construction in Liberia and finally responsible for the arduous task), referring to the ethnic mix employed in the project work force: "Rivalry disappeared and became growing buildings. This cultural mix again proves that the basic nature of men is one and the same, and when joined in a common effort they are able to subordinate ideas to the achievement of a common goal."

The influence of Charles and Shirley Leavell, and what they built, went beyond the hundreds of thousands of lives touched and the furtherance of improved lives facilitated by their medical and educational building. Their work on outer space projects may have helped save millions, and possibly even billion of lives.

C.H. Leavell and Company joined in a venture of world-altering events during the Cold War. The whole earth as we know it was in jeopardy of being destroyed in a chaos of flame, radiation, and horrible death, if not total annihilation of every living thing. It is most likely that the average world citizen, during all those years of tension and uncertainty about future existence, thought mostly of the ghastly warheads and the missiles that might deliver their personal vaporization. The missiles were the most obvious emblems of nuclear fear, but these instruments of destruction, on both sides, had to be assembled, dug in, encased in steel and concrete enclosures and mounted on massive pads. These had to be built with an exactness that controlled the fate of the world. Without these massive delivery systems we could never have won the Cold War—if anyone ever really wins such madness. Nevertheless, Charles and his company had to be thrilled underneath the delicate and deadly work. This was the Cold War that nearly everyone on this earth believed could become "hot" at any second.

The Leavell Company entered these critical endeavors at high financial risk. There were tight time schedules, extreme secrecy, and absolute maximum safety hazards. It was obvious that their experience with early military contracts and missile work would help, but now they were entering the ultimate game, where everyone could literally lose the world. There had never been anything near the likes of the Minuteman, Titan, Atlas, Saturn, Apollo and allied facilities. Charles and his company joined in allotting company money and skills with an elite group of construction companies to bid on and build these mighty projects: Morrison-Knudsen, Perini, Paul Hardeman, Johnson Drake and Piper, Peter Kiewit, Fishback and Moore, Scott Company. His friend Bill Scott—of the latter company—who had accompanied Charles on his trip to China, was especially essential in doing the most complex mechanical work. Their relationship had been nourished back on some Los Alamos laboratory jobs and now it flourished.

This was the greatest horse race of all time, with the wager the fate of the world and all its inhabitants. The owners had to pass word on to the trainers, to pass on to the jockeys, to whip the horse some if needed to cross the finish line first.

Very soon after these construction jobs began, they were slowed by the eternal problems of bureaucratic red tape. Charles helped save the day for himself and other contractors who were losing their Levis on fixed-price contracts. These documents allowed no financial flexibility for "change orders." So, in February 1961, Charles appeared as a witness before the U.S. House of Representatives' subcommittee on military appropriations. He, joined by other contractors, unsparingly stated their two main points of possible time and financial failure, winning the two things vital to keeping the missile site program on track, namely:

1. Direct the Air Force to honor change orders.
2. Appropriate the funds to pay for them.

After that, the consortium of firms moved out, stretching the harness to the limit and most of the participants becoming lifelong friends, because they were working together in a driving effort of critical minutes, hours and weeks to possibly save the world. No greater impetus or inspiration could have been imagined or needed.

The Leavell Company was heavily involved in the $272 million worth of Minuteman facilities in North Dakota, Montana, Missouri and Wyoming. Charles sponsored and joint-ventured, with Fishback and Moore and Kiewit, the headquarters structures at NASA's Manned Spacecraft Center in Houston, Texas; the Titan III launch facilities at Cape Canaveral and the Saturn rocket testing facilities. These were just a few of the major projects Leavell joint-ventured or sponsored during this time.

There is little doubt, looking back after the passage of years, that the Titan missile facilities were *the* benchmark in the company's project record. It wasn't just money and under- or overbudget schedule worries here. Far more was at stake.

The company was the sponsoring contractor on the South Dakota Titan I and then on the Titan III launch facilities at Cape Canaveral. The $54.8 million South Dakota job was joint-ventured with M.K. and Scott Company.

It was an awesome challenge. They had to move 2 million cubic yards of earth, pour 100,000 cubic yards of concrete and insert 35,000 tons of reinforcing steel. Each of the fifty-acre sites featured igloo-shaped missile silos, each constructed to accommodate the Titan missile, its launcher and a crib. The 160-foot subterranean silos were connected to propellant and equipment terminals, power and control centers and two smaller, additional silos fitted with retractable antennae. At completion, access tunnels linked the missile's complex elements. Then everything was backfilled, leaving only hatches visible on the prairie floor. As if this wasn't complex and difficult enough to deliver to exact measurements, the behind-the-scenes activities added to the daunting endeavor.

Immense amounts of legal knowhow were needed, and the projects consumed endless hours of work for the Leavell Company's law firm of Kemp, Smith, Duncan and Hammond, as they tried to assure that construction continued at the required lightning speed without getting lost in the colossal claims that inevitably arose. They had to overcome bureaucratic idiocy that allowed some ignorant, paper-pushing government employee of limited vision to risk the fate of the world by being officious over three rolls of paper towels. The endless stupidity of rigidity.

Then there was the Titan III project at Cape Canaveral—a totally reverse construction pitfall. The launch pads were above ground on watery soil. The dewatering; the near-instantaneous pouring of huge concrete fills and bastions; the massive pilings that had to be driven; a sixteen-story umbilical tower, including an elevator and thirteen retractable service platforms: these reveal only a part of the massive problem to be overcome. And Charles and company did.

The final results were a touch of unsung glory. They had given the Air Force its *first* complete launch system designed for military purposes, ready to be placed in a vertical or launch position on the pad for the fatal—or saving—countdown. Charles and Shirley were sitting in the viewing stand with Army Corps of Engineers and Air Force brass as the very first countdown proceeded. All the Leavell Company personnel were anxiously waiting at ground level. At last the liftoff, which generated more than two million pounds of thrust, came. The

Titan III missile facilities, Cape Canaveral, Florida.

Leavells and everyone else in the stands sighed, feeling great relief, some trepidation as to what the final results of this day might be and a natural exultation over a great success as they followed the missile heavenward with their eyes. But the people on the ground still had their eyes glued to the vapor residue slowly clearing from the pad. Many were yelling and waving their arms in jubilation, shouting, "It's still there."

Thirty long years earlier, they had built the pads that launched the first V-2 rocket soaring across White Sands, New Mexico, sending—along with their own hopes—the United States into the space age.

In the mid-1960s, near New Orleans, the Leavell Company and Peter Kiewit Sons and Company played a major role in the Apollo program to put men and equipment on the moon from Cape Canaveral. Before that, they undertook a different kind of national defense project, which certainly helped break any conceivable monotony: they built the barracks for Lackland Air Force Base at San Antonio. Scheduled as a breakneck project, these were more than just barracks; a local paper called them "superbarracks." They consisted of five self-contained, air-conditioned training complexes luxuriously appointed with complete kitchens, dining rooms and classrooms, with ten additional, two-story sleeping wings that comfortably held 1,040 men each. When a magazine writer visited the job site to do a story near the June 1969 completion date, a sergeant instructor said in a jest that was mostly truth, "We have everything here but mother." It made history for the Air Force, enabling them to speed up vital training exercises with maximum results. Another huge Leavell contribution far beyond the bounds of local thinking.

The Leavell Company had already helped make more space-age history at NASA's Manned Spacecraft Center at Clear Lake near Houston. The abbreviated details still show the immense effort involved. In thirteen months, they finished thirteen buildings for the nation's space program, maintaining the schedule despite 159 change orders. The nerve center of the nine-story project was designed to serve as control and support center for the entire NASA complex. One of its first missions was expediting the Gemini 4 flight by astronauts White and McDivitt in June 1965.

One can now more easily comprehend Charles H. Leavell's craving and need for the sanctity of his precious 4UR dream. The mental visions he had of its healing spirit, of its mountains and vast blue sky . . . well, the mind's eye relaxed the soul at just a glance. The eternally talking waters of Goose Creek—the flowing, rippling castle of the rainbows, browns, and cutthroats of its tributaries—both excited and soothed his weary brain at just a thought.

Yes, the Leavells deserved this place of peace, just as they contributed so very much to world peace all the way to the falling of the Berlin Wall. Even as that mind-bending, world-altering event took place, the trout of Goose Creek delightfully teased and pleased the Leavells of El Paso and other environs.

CHAPTER 15

The Wiles of Walton
and the Last Trail Winding

There is little doubt the Utes thought of the 4UR hot springs and the surrounding lush hunting as sacred. Even those who initially took it away from them felt its magnetic and healing pull. The first visitors to the springs after the Utes could not help but notice their long-worn trails to this special place.

The 4UR was at first called the Wagon Wheel Gap Ranch. The discovery of an old wagon wheel gave it the name, though no one knows for sure how the wheel got there. Many thought it was part of the disastrous John C. Fremont expedition in 1848. Others said it had been left after a wreck by Kit Carson. Still others think it was from the remains of an Indian raid on either American or Spanish explorers.

The Leavells changed the name "Wagon Wheel Gap Ranch" to 4UR. They took the two Ws and turned them into four Us and added the R for ranch. They also registered 4UR as their brand.

In 1872 the first claim to 140 4UR acres was acquired by four settlers, but it was a group of successful mining men from Lake City, Colorado—Charles Goodwin, Henry Henson and Albert Mead—who bought it and started the development of the hot springs. At that time, the Goose Creek hot springs had more than thirty pools that graded in temperature from near ice-cold to actually bubbling hot water.

These far-sighted men anticipated developing the springs' curative powers for arthritic miners and other incapacitated persons, as well as a comfortable hotel for any who chose to visit this spot. They would never know how many noted and powerful people would pick up on their dream and move it forward all the way to the present with Charles H. Leavell and his family.

The hot springs hotel did become a favorite location for nature and outdoor sporting enthusiasts by the late 1800s. Reports from 1889 reveal that between four and five thousand pounds of trout were being caught each season on the Rio Grande next to Wagon Wheel Gap, now part of the 4UR. What wasn't reported was the tons and tons that were dynamite "caught" and used to feed hungry miners.

Nevertheless, author Ernest Ingersoll wrote, after visiting the springs in the 1880s, "I know of no place in Colorado where the fly fisher will have better sport, and the angler though uninstructed in the wiles of Walton, get better results. The area offers geologizing, botanizing, and general natural history enough to invite study in endless variety." Ingersoll also praised the springs' curative powers: "There are accounts of men brought here utterly helpless and full of agony from inflammatory rheumatism or neuralgia who in a week were able to walk about and help themselves . . . and who in a month went back to work."

It was obvious even then that the springs at the ranch were close to sacred to these explosion-jarred, dust-breathing miners, who sometimes slaved worse than rats deep underground, often damp or wet for month after month, stooped and bent in every joint. All their blasting, digging and mucking, in this body-breaking, life-endangering and extraordinary sacrifice, went for many comforts and wealth to others.

It was just as obvious that the elite wealthy could find succor here, as well as great fly fishing and hunting adventures. General Palmer had extended his Denver and Rio Grande Railroad to the ranch and other commercial and tourist points by 1883. The railroad depot he built is still there on the banks of the Rio Grande, cleverly turned into a captivating private home today. The railroad replaced the old stagecoach line and put it out of business.

General Palmer purchased a controlling interest in the ranch in 1875 and under the management of Job C. McClelland started upgrading the hotel and springs. He kept the cost of a pool soak to $.50; a private tub was $.75, salt glow was $.75, a bathing suit $.25 and an

The Denver and Rio Grande depot at Wagon Wheel Gap. (Courtesy, Denver Public Library, Western History Collection)

alcohol rub the same. A complete body massage was $1.00. In that first year it was reported that over one thousand people visited the springs. In 1902, at great expense, Palmer financed the concrete bathhouse, which contained two large, hot, sulphur swimming pools, and added bedrooms with private hot baths and individual tub and spray baths. That same year a reservoir was built to ensure an ample supply of pure water; ice was harvested from this reservoir in the winter and was put in insulated storage for summer use. In 1904, the St Louis International Exposition awarded the ranch springs a silver medal.

The daily train service to Wagon Wheel Gap continued, and in 1904 a Mr. Bergay began management duties. By 1906, the improvements at the springs had reached over $100,000 and General Palmer announced that a fish hatchery was to be built. The general was delighted at the help the springs were giving injured and bruised miners, but he was also pleased that the warm waters offered entertainment and contentment to royal and fashionable dignitaries from all over the world, including his military friends from his old regiment. The ranch was giving the unselfish general himself immeasurable pleasure. He loved to see others enjoy themselves, regardless of rank or color. Reports that Palmer urged that the hot springs be heavily promoted as a resort are probably quite accurate.

Among his seemingly endless improvements was a sixteen-foot-wide road up Goose Creek that he called a "boulevard to the beaver lakes," where close to 25,000 fish were stocked. It was surely a wonderfully chaotic time of both construction and destruction during Palmer's final years in Colorado. In spite of some health setbacks, he carried forward with the fine emotions of *giving* that this enchanted place spawns in all who are kind and understanding of this precious land, its water and its wildlife.

One particular of the time demonstrates Palmer's truly great heart. Immediately after the turn of the century, negotiations were started for the sale of both the Denver and Rio Grande and the Rio Grande Western railroads. Palmer had not run the firm since 1883 but still retained the largest holding of stock. When the firms he had founded were finally sold to the world-famous eastern financier, Jay Gould, Palmer's take was $15 million. This was a few million more than he had anticipated, so at last, in his mid-60s, he could fulfill the final part of the dream that had come to him on a train in 1870. Without hesitation, he made one last journey of destiny fulfilled, over the hundreds of miles of tracks he had first pioneered, through the majestic Rockies, in his favorite private car "Ballyclare." He handed out the hefty profits to every railroad employee with quiet dedication and joy. No matter how long or how short their employment, he gave them the money—whether gandy dancer or ticket clerk, they received his bounty. One long-time ticket agent in Utah was given $35,000. Most of the gifts

General Palmer allowing the help to pose in his luxury touring car in front of the Garden of the Gods. (Courtesy, Western History Department, Penrose Public Library, Colorado Springs, Colorado)

averaged about $5,000 per person—an astounding amount in a time when $5,000 would buy and furnish a five-room home or start a business with capital to spare.

In spite of all his wealth and multitudes of loyal friends, Palmer knew sadness as well. In 1880 his beloved Queen suffered the worst of what was to become a series of heart attacks. Her heart could not take Colorado's high altitudes. In 1883, the year he was building the railroad depot at Wagon Wheel Gap, she and their three daughters, Elsie, Dorothy and Marjory, moved to New York. In 1885 they moved to England.

It is not hard to imagine that, in spite of his love for Glen Eyrie and Colorado Springs, the Wagon Wheel Gap Ranch became tremendously important to Palmer in both body and spirit. Twice a year he crossed the great ocean to visit, console and give love to his family in England. In 1894, just before Christmas, Queen passed on. She was only forty-three years old, but they had been man and wife for twenty-five years. The shattered Palmer eventually recovered, as he always had, and brought his daughters back to Glen Eyrie, where the mighty wooden mansion had been converted to a stone castle, to finish growing up.

During these supposed retirement years, Palmer was constantly involved in national affairs, as well as those of the West and Mexico. Even though he had been a heroic Civil War general, he was still a Quaker. He abhorred the anti-Spanish propaganda that appeared prior to Teddy Roosevelt's charge at San Juan. However, when the Spanish-American War was over and evidence of scandal and blundering in the military were revealed, he threw his entire weight behind efforts to uncover the misdeeds.

In spite of Palmer's efforts, Teddy Roosevelt became a national hero. When the new hero came to Colorado Springs to celebrate the anniversary of Colorado's statehood, he spoke in a park at one end of town while Palmer expounded from the porch of the Antlers Hotel at the other. Palmer drew a larger crowd than the then-vice president.

The general was asked to run for governor of Colorado in 1902, but never seriously considered it, even though most believed that he would have been a sure winner. No one in the mountain state was more loved and respected at the time. On this point historians are in full agreement. The truth is, Palmer disliked politics and speaking in public.

His many continuing projects included making blissful trips to Goose Creek; funding colleges and universities; and giving large gifts of cash or land to libraries, sanitariums, Colorado's first forestry college and a boarding school for Negro children in the south. One of Palmer's biographers maintains that he kept scores of people on the payroll simply because they needed the jobs. He was a brilliant businessman, but also a sucker for hard-luck stories.

While enjoying a gallop through the mountains with one of his daughters and a friend on October 27, 1906, riding one of his favorite horses, Palmer's mount stumbled, throwing him hard against the very

The first Antlers Hotel, Colorado Springs, Colorado. (Courtesy, Western History Department, Penrose Public Library, Colorado Springs, Colorado)

earth he had fought to preserve. His neck was broken and he was permanently paralyzed from the waist down. It was not long, however, until he ordered a special steam car built, and had himself driven through the mountains, again accompanied in the huge car by a doctor, nurses, children and dogs. He insisted on going up mountain roads that chilled the spine of the driver.

The following summer, in August 1907, the general tossed the biggest celebration Colorado Springs had ever seen. For years he had gone back to Pennsylvania for the annual reunion of his old regiment, the Fifteenth Pennsylvania Cavalry. But now, being paralyzed, he decided to bring "the boys"—now old men—to him. They came.

From the time they left their homes until they started their return two weeks later, every single cost was picked up by Palmer. He reserved the entire Antlers Hotel (a place of luxury in those days) for his men. And did they celebrate! They toured the mountains and fished Goose Creek. They saw Colorado's majesty from special trains. One evening all 280 of them feasted at Glen Eyrie.

One of General Palmer's beloved horses. (Courtesy, Western History Department, Penrose Public Library, Colorado Springs, Colorado)

Some of his men stretched a big banner in the Antlers reproducing the famous order the general had been given, to get that "son of a bitch Jeff Davis dead or alive." Palmer, who had pursued Davis uncountable miles before one of Palmer's brigades captured him, had the banner removed because one of Davis's daughters lived in Colorado and they had become friends. Yet another entwining of the constantly crossing trails of history.

In June 1908, accompanied by all needed doctors and staff, Palmer sailed for England for the marriage of his daughter. The wedding was called off but he stayed for a few months, even though he had to be carried on a litter like one of his wounded soldiers. His body weakened but his mind stayed spry and indomitable.

He returned home to Colorado and lived several more months, but by March he fell into a coma. He died on March 13, 1909. At the probate hearing it was discovered that he had given a great deal of his fortune to his laborers, friends and charities. Even so, there was plenty left to take care of his family.

Palmer's physical legacy is all over Colorado, and a great many other places of the world, but his truest and greatest legacy was his attitude toward life.

He had also pioneered Charles Leavell's great dream—the place of peace now soon to be forever named the 4UR.

CHAPTER 16

Buying Up the World and Hoping for Paradise

Amidst all the world traveling and construction, Charles was so involved with dreams of the waters of Goose Creek that it had become the very blood stirring in his veins. He and Shirley started buying other ranches, searching and hoping for a suitable substitute until the "dreamed-of someday" when the 4UR would be for sale. The other ranches did help temporarily—like aspirin eases a toothache until its effect wears off. Nevertheless, Charles went at it with the same great zest as he did everything, big or small, in life.

All the ranches, no matter what their acreage, had been host to events of historical significance throughout western history. There were nine in all that he bought, owned, traded for or inherited. The first one, of course, was the childhood ranch he had loved so much and was forced to sell to pay a debt: the Figure 2, north of Sierra Blanca, Texas, in the Delaware Mountains. Then came Rancho Felix, located at Vinton, Texas, which he acquired in 1940 and sold in 1949 after he had seen the 4UR. The Spanish Peaks Ranch, ten thousand acres on the northwest slope of the Spanish Peaks west of Walsenburg, Colorado, and southeast of La Veta, Colorado, was where so many of our country's early explorers rode and trod. The 3R Ranch, purchased in 1963 from J. D. Leftwich, another contractor, was located near Rye, Colorado, also in the 4UR realm. It was twenty-three thousand well-grassed, well-watered acres, with a carrying capacity of thirteen hundred mother cows.

The historic Wine Glass Ranch—all on the Purgatory River northeast of Trinidad, Colorado, near Bent's Fort—had seventy-three thousand acres with a carrying capacity of one thousand head of mother cows. That ranch had been the site of many Indian battles. The Leavells once found a weathered old stagecoach that had been abandoned for unknown reasons; they also found dinosaur tracks on a shale rock at the bottom of an arroyo. In a tragic addition to the ancient and old West history here, their ranch foreman was drowned and smashed to death when a flash flood caught him riding his horse down a dry gulch. Charles purchased this ranch from the Rourke Cattle Company in July of 1971 and sold it in 1974.

The Van Horn Ranch, located five miles east of Van Horn, Texas, and often referred to as the Cockrell Ranch, was purchased in early 1970. Charles was a partner in this one with his son-in-law, David R. Winton, and Marvin Porter and T. W. Winters. The latter two were big cattlemen from Bryan, Texas. They drilled and struck good water, adjudicated the wells and subsequently sold the land to Albert Ivy, principally a farmer of cotton, vegetables and pecans.

The River Bend Ranch, near Del Norte, Colorado, at the time of purchase, consisted of 580 irrigated acres with 1.5 miles of frontage on the Rio Grande. Then Leavell and his daughter, Mary Lee, purchased 150 head of registered longhorn cattle, so they needed more winter range. In 1982 Charles traded with the State National Bank of El Paso for two adjoining ranches and gave all three the name "River Bend Ranch." The River Bend Ranch was in magnificent meadowlands and it wasn't far to the San Juan range, where they summered the longhorns, bringing the animals down to the lower country to feed on the lush meadow grass.

One observance impressed Charles mightily: the fact that a wild longhorn bull would stand guard over a cow while she calved. This was unheard-of in other breeds he had handled. Even though they experienced some difficulties beyond the ordinary, Mary Lee took to the study of the longhorn breed with perseverance and dedication. Her reputation in this field grew swiftly and soon she was elected to the board of directors of the Longhorn Breeder's Association of America.

Once, while Mary Lee was in charge—Charles and Shirley were vacationing in Hawaii—she called and told her father that she had just been to an auction where she bought a fine breeding bull from the King Ranch National Wildlife Refuge in Oklahoma. Charles was pleased when his daughter boasted that the bull, named Prairie Fire, was one of the best in the world and would improve their herd a lot. However, when she told him the price she had paid for the animal he almost dropped through his shoe soles—$48,000 was a little more than he expected them to bid on a breeding bull!

The Leavells returned from Hawaii in time to be there when the prize bull was delivered at the River Bend Ranch. As it arrived in a carefully padded trailer and was unloaded, Charles had to admit that the tricolored, blue, brown and white animal was majestic. They turned the bull loose at the water tank. The great Prairie Fire drank, then turned, smelling the air. His lips and nostrils curled up as he caught the scent of some heifers in estrus in the meadow just below the water tank. He started toward them, bawling low in his chest. Charles, Mary Lee and the River Bend crew all watched proudly as the expensive bull appeared to be delivering on their investment without delay.

Finally he mounted a heifer—and just as everyone breathed in relief at apparent fulfillment of the promise of many valuable calves to be sired by Prairie Fire, he let out a half bawl, half scream, falling off the heifer and down on his side. They called a veterinarian and raced down to wait helplessly by the paralyzed bull. They were stunned at the verdict when it came: Prairie Fire had pursued his calling with such energy that he had broken his back on the very first try. The heifer was finally bred by another, less expensive, but more durable, bull. Prairie Fire's hide was tanned and brought $500.

As renowned as Mary Lee had become in the longhorn business, a few incidents such as this do become disenchanting, and the Leavells had far more than their share. The River Bend Ranch was not sold until 1994, although the longhorn business had been discontinued earlier.

Charles Leavell was now in the habit of ranch adventuring while profiting at the same time. This seemed to be a natural talent, and he certainly developed it to a high degree over his entire life.

In 1972 Charles purchased the Three Rivers Ranch, near Tularosa, New Mexico, from Thomas Fortune Ryan III. The Three Rivers played an important part in New Mexican and Southwest history. Adjacent to the Mescalero Apache Indian Reservation to the east and the national forest to the north, it consisted of fifty-three thousand acres. It ran six hundred mother cows, partly because it almost always had water and fine grazing from the melting snow of White Mountain. This mountain, held sacred by the Apaches, dominates this vast area, just as the mountain (Sierra Blanca) near Alamosa, Colorado, dominates the massive, varied terrain of the southeast part of the 4UR realm.

The actual dealing on this spread was a little bit delicate and yet very straightforward—just like both the men involved. Before Charles agreed to the purchase, he and his pilot flew it out several times. Besides the abundant water in streams, springs and tanks, the land was also rich in wildlife. They spotted bear, deer and all sorts of small game, as well as elk in the adjoining national forest.

At the headquarters Charles found two appealingly landscaped acres surrounding a beautiful four-bedroom hacienda with a pool and bathhouse, a fine patio, a green house and a trap-shooting range. There were other well-made buildings for the help, with good barns and workshops. At the nearby hunting lodge was a four-thousand-foot landing strip of gravel and dirt, and another had been built at the main house, called the Ryan House. It even temporarily crossed his mind that this might possibly be as good as the 4UR. All in all, a sort of paradise to a man and family who loved business adventures, hunting, fishing and flying—but paradises can often conceal both passion and poison.

Charles describes the beginning and the end this way. "Thomas Fortune Ryan asked us to come up one Sunday and have a picnic with them. Oh, they had everything to eat and drink, all kinds of meats, vegetables cobblers and some good wine. After awhile he looked off at the great white mountain and said softly to me—I don't think anyone else heard—'Do you want to buy my ranch—' I asked what he wanted for it. He said, 'I want a $100 an acre.' Well, I looked across all that beauty of hills, grass, mountains and I could see in my mind the game running and flying free and the cattle grazing fat and contented. I said, 'Okay, I think you got a deal.'

"Oh, it was beautiful with the three rivers coming down from Mount Blanca near Ruidoso and all the other things I loved were here. But after we bought it I found that in all this beauty we really had some troubles. James H. Polk III managed the ranch from our offices in El Paso. He was the son of General Polk, commander of U.S. Forces in Europe.

"One year it didn't snow or rain enough to fill a sewing thimble. We ran out of grass fast and I was bringing hay up there by the truckload for those dadgum cows. All of a sudden a little moisture came down and out popped the locoweed. The cattle were eating it like dope fiends and were stumbling around like a bunch of drunks at a barn dance. I never, ever saw anything like it. The cows' tongues were hanging down like neckties and we lost at least a quarter of the entire herd before we could get them properly inoculated. This is a rare event, calling for weather of an exact kind so that the weed is the only green thing to eat and the hungry cattle go wild for it and crazy afterwards. Well, I didn't want to wait around for such a colorful event again so that's when I decided to start looking for a buyer. With all its history, I'm surprised even today I sold it. I must have inadvertently eaten some locoweed myself."

The main town near the mountain is the resort of Ruidoso, home of Ruidoso Downs and the world's richest quarter-horse race. The worshipped mountain certainly has returned the love and respect of the Mescalero Apaches. They operate their own ski lift there and run a carefully controlled commercial big-game hunting and fishing operation; all this under their indomitable Chief Wendell Chino. He also saw to the financing and running of the Mountain of the Gods Inn, with a golf course, fishing lake and gambling casino.

In another strange interweaving of historical figures, Charles bid on and won a construction job for the Mescalero Apaches, who count among their many noted ancestors the mighty warrior, Geronimo. Thus Charles Leavell's company ended up building the Mountain of the Gods Inn, the golf course, the lake—the entire complex. At the dedication ceremonies, Charles was persuaded to give the main speech. Other notables were in attendance, but their presence was overshadowed by the powerfully-willed Chief Chino and his beautiful Cherokee wife.

Charles was just getting his speech going when he spotted a really tall Indian standing at the back of the crowd. The man was clad in a big western hat and high-topped cowboy boots. That's all. Just as Charles spotted him, the naked man "streaked" through the stunned crowd and vanished somewhere in the adjoining turmoil. Charles's speech also vanished at this instant.

The Three Rivers Ranch was aptly described by Virginia Chappell in May 1956, in the *El Paso Times:*

> Friendly ways of the old West have never died at world-famous Three Rivers ranch—whose origin was the outgrowth of a dream by a big Irishman, named Pat Coghlan, then became the passion of Albert Bacon Fall, and now is a favorite treasure of Thomas Fortune Ryan III. . . .
>
> The silent, sun-drenched land, sweeping westward from the foot of serene Sierra Blanca peak in the White Mountains always has entranced its owners. The great ranch once included one million acres.
>
> Coghlan secured a contract for selling beef from his Three Rivers herds to the army post at Fort Stanton. In the meantime, he had engaged in a deal with Billy the Kid, who was rustling cattle from Texas up into New Mexico. The L X, a big Texas Panhandle ranch, had been losing a lot of cattle to rustlers. In 1881 Charlie Siringo, noted cowboy detective, found a number of hides, bearing the L X brand, hanging on the corral fence at Coghlan's ranch. Later, the Irishman was arrested and indicted by the grand jury at Mesilla, on a change of venue.
>
> Mr. and Mrs. George Nesmith, who worked for Coghlan at his ranch, had given information against him to Siringo. They were subpoenaed by the prosecution for Coghlan's trial, but were murdered along the White Sands, while en route from Three Rivers to Mesilla.
>
> Coghlan escaped the cattle charges with light fines, but soon was confronted with a more serious one. It was learned that two men, named Maximo Apodaca and Rupert Lara, had killed the Nesmiths. The two men claimed that Coghlan had offered them $1,000 to do the job. But when the killers came to trial, and could not identify the Irishman as the man who had hired them, Coghlan was absolved of implication in the murder of the Nesmiths.
>
> Coghlan's trials had cost him a lot of money and he had to begin getting loans—interest rates, at that time, were extremely high. His many promissory notes were to ruin him financially. In July, 1906,

Coghlan had had to give his Three Rivers Ranch as security for payment of a large note and its accumulated interest. He was unable to meet the payments. On January 28, 1908, Coghlan signed a warranty deed to Albert Bacon Fall, who had bought the note originally held on the ranch by Reymond.

The Three Rivers Ranch became historically known as the A. B. Fall Ranch after its owner Senator Albert B. Falls, who had resided at Hillsboro, New Mexico, while getting three horses in the Kentucky Derby. He was both nationally famous and infamous for many other activities.

One of Fall's first ranches was the 3R, previously mentioned. Albert Fall practiced law in a house he kept in El Paso. However, his Three Rivers ranch dominated everything he owned. As he steadily increased his herds, he acquired more rangeland. His dreams of creating a ranch and cattle kingdom were reinforced greatly when H.P. Everbart, a son-in-law, and the Thatcher brothers formed a consortium in Pueblo, Colorado, and bought the Bar W, thus controlling nine hundred thousand acres of land south and west of Carizozo. Before long, with the help of his son-in-law, the Fall and Thatcher ranch holdings merged. Soon they purchased the large Harris-Brownfield ranch holdings nearby. All this property had a colorful historic background.

Albert Bacon Fall's next purchase was from a woman known as the cattle queen of New Mexico, Mrs. Susan McSween Barber. She was nearly as tall as Fall, had dark blond hair, and she could ride and shoot as well as any of them. Alexander McSween, her first husband, had been killed in the Lincoln County War.

In his long-sought-after domain, Fall controlled over one million acres with thousands of white-faced Hereford cattle grazing across it. Fall's son, Jack, was in charge of all undertakings of the vast empire, so it gave Fall plenty of time to increase his political activities. He had changed over from the Democratic Party to the Republican and was elected by New Mexico's first legislature, in 1912, as one of the first two U. S. senators from the state. He was soon to become Secretary of Interior under President Warren G. Harding.

Then disaster and disgrace overtook him. First, he lost his son, Jack, during the deadly 1919 influenza epidemic. This was a heavy blow. Scandal followed grief.

Somehow, a company headed up by Fall's old buddies, Edward L. Doheny and H. F. Sinclair (good friends from their Hillsboro/Kingston, New Mexico, mining ventures), was granted the rights to some rich oil-bearing public lands on vast United States Naval reserves. The senator was accused of bribery in the letting of the lease rights. Word spread rapidly before the trial that unknown hands had passed $400,000 into Fall's pockets. Most everyone was aware that he needed funds to advance the Three Rivers Ranch.

Albert Fall spent seven years of costly and intense struggle in the courts. He was found guilty of bribery. He got a prison sentence of a year and a day and was fined $100,000. Doheny was tried on the same charges, the same evidence and the same witnesses, but was acquitted. Now, over seventy years old, he sought out peace for a few months on the ranch before being incarcerated in the New Mexico state penitentiary as a federal prisoner on July 21, 1931. He was released on May 9, 1932.

His wife died in 1943 after trying desperately, and in vain, for years to restore her husband's good name and save at least some of his property. Soon after his eighty-third birthday, Fall himself died on November 30, 1944, in Hotel Dieu in El Paso. The heavy fines and other associated legal expenses had forced him to sell his beloved Three Rivers Ranch sometime earlier.

The ranch was subsequently sold and resold over the years. Its acreage was near the ranch (previously mentioned) owned by the pioneer ranching Cox family, which became the firing ground for the world's first atomic bomb, at a facility where Charles H. Leavell was to play such an important role. The history of the world was altered forever there, in an explosion of sunlight-without-sun instant. In the final history of the earth, this melted spot near the Three Rivers Ranch might be the most famous of all.

Rob Cox carried on the history-making in this area as a hard-working founding member of the Farm and Ranch Heritage Institute Museum, sited on New Mexico State University land at Las Cruces,

New Mexico. Its opening, in December 1996, shows its strong promise of becoming the greatest agricultural museum in the United States and eventually the world.

When the Three Rivers country was purchased by Thomas Fortune Ryan III, it included 110 sections. Ryan came from a line of nationally known financiers. His grandfather was a Wall Street broker in tobacco. His father, Thomas Fortune Ryan II, built the railroad from Chihuahua, where Thomas Fortune Ryan III was born.

Ms. Chappell also described the vast improvements made by the Ryans.

> In 1941, he built an elegant new home of Spanish architecture under the now towering pecan and poplar trees set out by Albert Bacon Fall. At the west of the house, which provides an unmarred view of Sierra Blanca peak, is a private swimming pool and dressing rooms. Beyond the pool, is a greenhouse and Ryan's personal trap-shooting grounds.
>
> Ryan converted the old Fall mansion into comfortable quarters for his farm foreman, S.M. Cozzen, who has been with him for many years. Not far from where Susan McSween Barber's stone house stood, he built a handsome Spanish-type home for his ranch foreman, Carroll Johnson, prominent in ranching circles in the area for the last 21 years.

Right after World War II, Ryan built a fine hunting lodge, named Bataan Lodge, in the White Mountains on the extreme northeastern boundary of his property, specifically for the relaxation and entertainment of veterans from that infamous death march. He entertained there often during big-game hunting seasons. He and the Lions Club of Tularosa created a Billy the Kid Rodeo, which has become one of the outstanding events for young rodeo performers in New Mexico, featuring a day-long celebration ending with a big barbecue on Ryan's Three Rivers Ranch.

Ms. Chappell finished,

> Tom Ryan also is a Roman Catholic. Soon after he acquired his Three Rivers ranch, he began the restoration and refurnishing of a small Catholic church, built on the property by Fall. It is located about 10 miles from Ryan's own ranch home, and is across the road from ruins

of an old adobe building, where Billy the Kid often hid out when escaping the law. The church is named Santa Nino and here, each New Year, Tom Ryan gives a feast for those who attend the parish and numerous friends, including many Mescalero Apaches.

Thomas Fortune Ryan III calls his ranch "home." He says he never will sell this fabulous ranch, whose history began in 1874, when an ambitious Irishman, named pat Coghlan, invaded the Three Rivers country.

And he didn't sell it until after he had put many miles, many years and much love into one of the most significant ranches in America. But alas, all things change. Ryan sold it to Charles H. Leavell, who also loved the ranch until nature took control and he sold it, in April 1978, to a famous rodeo hand named, R. E. (Sonny) Wright. The time of the Leavells' ownership only proved how the paths of great pioneer men and women the world around are intermingled on occasion. Yet an event of nature beyond his control, and beyond all the ranch's history—the locoweed—forced Leavell as well to sell.

Except for the 4UR, by 1978 Charles Leavell's principal days in ranch dealing were just about completed, albeit in high style. He had broken even on one ranch and lost money on one, but over all he had had a hell of a good-time ride and made lots of money doing it.

There is still one other ranch—the most important ranch besides the 4UR—in Charles's trading, and this one certainly created one of the most adventurous stories that could be conjured up in the West. As well as becoming very personal to Charles and Shirley, and being forever imprinted on the lives of Pete and Mary Lee, this ranch would also be indelible in the minds of Mary Lee's children.

It is the 115,000-acre J M Ranch between Gardner and Westcliffe, Colorado. All of it lies in the Wet Mountain Valley that Zebulon Pike had crossed. When Charles bought it in 1961, the land had a carrying capacity of twelve hundred mother cows.

Charles explains the initial business side this way: "Three ranchers, all neighbors, were quarreling over fences, grazing and water rights. I made all three an offer of 7 come 11, $711,000, and they made friends quickly, resolved their differences and took my offer.

JM country.

"I subsequently bought five other neighboring properties. The ranch was on top of the Sangre de Cristo mountains, encompassing the Wet Mountain Valley. We built our own private airstrip; developed roads, ditches, fencing; and helped make the entire area magnificent hunting of elk, deer and antelope. We improved several streams and small reservoirs for fishing and better and better vegetation growth. There were also coyote, bobcat, mountain lion and wild turkey. We struggled, mostly with success, to put this grand stretch of ponderosa pines, streams, hills and valleys, meadows and mountains into natural balance. We did more work here than I have breath to tell, but it was all worth it."

Right after closing the first purchase, Charles hired a top ranch manager named George Baker, who was supervised by Joe Lea out of the El Paso office. Baker now lives in Denver and is a well-to-do businessman. Later Charles would hire the black cowboy, Sherman Grundy—who had been such a part of Charles's youth on his father's ranch—and bring him up to the J M. A retired buffalo soldier, Sherman

JM Ranch, 1962. (Left to right) Charles Leavell, Joe Lea, David Winston, and George Baker.

knew just about everything there was to know about cows, horses and the land from which they lived.

Mary Lee had met and married David Winston while they were both finishing their studies at Stanford. He was from a prominent, wealthy family from Houston. To Charles's way of thinking, they needed a hard place to start and learn, so he asked George Baker if he could find some work on the ranch for his new son-in-law. George said, "Sure, send him on out and I'll find plenty for him to do."

Charles Leavell explains, "I knew that David had a little bit of ranching experience with his own family and he could ride good horses. Of course, Mary Lee had already excelled in every kind of riding from western to English show horses, where she won many ribbons, so she would be fine. So they all went at it and did a hell of a good job as I'd believed in the beginning."

It soon occurred to all concerned that Mary Lee and David needed a permanent housekeeper/cook. An impressive Mrs. Meeler was given a try-out by Baker. He told Charles, "I don't know anything about her except her person is as clean as a hound's tooth, a fine cook and knows all about children. And she's lived on a farm and ranch in Oklahoma."

So Charles said, "Go ahead and hire her. In the meantime I'll check her out. I don't want to be in any way derogatory or critical about the Denver police department, but they found her clean—no record at all.

"Mrs. Meeler was on the ranch five or six months, making a fine, dependable employee, when we shockingly found out back home in El Paso the terrifying truth about her. I forget whether it was myself or Shirley, it doesn't matter, but one Sunday we were reading the paper and there was Mrs. Meeler's picture right there in the Sunday supplement. She was one of the five most-wanted women in America by the FBI. We were scared silly for our children, but we went into action instantly. Shirley called the ranch, telling Mary Lee to get to a pay phone immediately, and call her back."

Mary Lee, being as bright as anyone else in a family of brilliance, gathered the children and raced and roared the pickup motor nearly twenty miles over a dirt road to get to a pay phone. The first thing Shirley wanted to know was the whereabouts of the children. When Mary Lee said they were right there with her, Shirley felt huge waves of warm relief wash over her entire being. Then she found out from Mary Lee that Mrs. Meeler had vanished the previous noon, so they notified the FBI and other law enforcement agencies. The Leavells were told that the police were already in full pursuit of Mrs. Meeler. No one ever knew how word had gotten to the fugitive that they were closing in on her. The FBI found and captured her at Texas Creek on the Arkansas River, about one hundred miles away, where she was hiding in a cabin. They had traced her telephone calls from the ranch to Texas Creek, where she had a friend or relative in the small town. She is still in the penitentiary today.

"She'd never killed anyone on record, but she had seven or eight armed robberies, automobile theft, and two jail breaks, and of all things, she's even been charged with rape," Charles says. "It's still frightening to talk about that woman because of our children. Although she'd never been convicted of a murder, there's always that first time." Charles

Longhorn cow #007.

figures that the reason she was so dedicated to her work at the ranch, and never caused any harm, was the simple fact that everyone there, especially the children, had treated her with great respect and kindness.

The Leavells and David Winston were able to run about two thousand head of mother cows because of wet years. The winters were rough in this summer/fall paradise, with deep snows and freezing winds. They had to do a lot of hay hauling and cake-cube feeding, sometimes through belly-deep drifts. Even though they had forty good cow horses, and it was a gut-real ranch operation—with the land running all the way over into the Great Sand Dunes National Park at the foot of the great white mountain in southeast Colorado—they all handled it as part of life's deal on a ranch. It was. All the other improvements to this often lush, often severe land did not just happen; a lot of bone-bending, back-tearing, hand-blistering labor went into them. But the country was so gorgeous, whether in summer green, fall gold, or winter white, that all hardships were simply disregarded.

Branding on the JM Ranch.

Charles and Shirley, as well as their friends Sarah and Tom Lea, spent a lot of time enjoying to the fullest the fishing, hunting, and just plain visiting in this land of beautiful contrasts. Charles and Tom rode and looked out over the vast arena of the wild in the same spots as Fremont and Pike, and countless Comanches and other Indian tribes had. There, without consciously knowing it, their souls and the spirits of their predecessors were meshing, melding, intermingling into one great timeless web.

The Leavells built several guest houses to care for visiting friends—locals as well as out-of-towners. Every fall they would have friends in from Washington, New York and other points on the globe. For those who hunted, there was an overabundance of game. There were two or three hundred antelope on the ranch (at which they always took running shots, to make things more even); during elk season nearly every hunter would get a big five-pointer. Everyone would feast and have fun. The elk were dressed out and the meat was either shipped back to the guests' homes or given to charities.

Charles Leavell with rainbow trout on the JM Ranch, 1962.

Tug-of-war at a JM Ranch Fourth of July picnic, 1963.

Every year on the Fourth of July they celebrated with a big picnic, inviting everyone around. They would shoot off wild displays of fireworks like little exploding comets in the vast clear sky. Charles says, "Oh, those were grand years of almost impossibly hard work and joyous play."

It was inevitable that after a time, David Winston, of such dedication and upbringing, would want to go into business for himself. He and Mary Lee had proven for years that they could handle hardships and complicated business dealings on the J M. They moved to El Paso to go on their own. David successfully invested in real estate and other enterprises, with Billy Sanders.

At any rate, Charles and Shirley gave the two young people their blessings. The J M had accomplished many things for the Leavell family. Besides all the pleasure, his children and grandchildren had been prepared for life through trial and intimate association with the power and beauty of nature; a blessed beginning indeed. However much all of this meant to Charles Leavell, he was to sell it in the same year he would finally get to make the bid on the 4UR, the ranch he had really wanted all along. Charles says, with many mixed adventurous thoughts and deep feeling in his voice and eyes, "My God, I loved that place with a passion, but it still wasn't the 4UR."

CHAPTER 17

The High and the Low:
A Necessary Love Affair

The San Luis Valley has a hundred-mile expanse of lush, fertile soil surrounded on three sides by massive mountains. It exists because of the bounty of irrigation water, and its abundance of timber, minerals, wild game and supreme fishing waters make it an all-around beautiful place just to breathe in and look at. The valley had for untold years offered its bounty of buffalo and antelope to the Utes and other tribes; during last two hundred years it has supplied many agricultural crops and way stations to hungry miners, cattle ranchers, tourists and the lovers of beauty who live there. The great valley and the mighty mountains were, and are, lovers who could not properly exist and reproduce without each other's fertility. At first, the valley was simply an extension, or prairie suburb, of historic Taos, New Mexico, but soon, with the Spanish settlements and later the incursion of white people, the valley and mountains became an entity independent of the Taos influence.

Most of the valley was divided into two counties. Costilla County absorbed the northern and eastern areas of the valley and Guadalupe County covered most of western side all the way north to the Rio Grande. The name of Guadalupe County was changed to Conejos County after only a week.

Transportation was always a major part of all settlement, and in the San Luis Valley that could not be properly done without crossings at the Rio Grande. Two prominent politicians and developers named

Francisco and Head installed two ferries across the Rio Grande, one for La Loma and another at a point near La Sauses. In that productive year a third permit was granted to a man named Garcia for a point at the mouth of the Culebra River where it met the Rio Grande. The place was called Paso del Norte in the early 1860s. The three transport spots were developed in expectation of droves of prospectors and settlers to come—which they surely did.

The rates seemed high but were actually about the same as other ferries in the territory. A buggy with one horse, mule or ox was $1.00; a wagon and team was $2.00; loose sheep, cattle or horses were charged $.05 or $.10 per head according to guessed size.

The ferry at La Sauses, which was called Stewart's Crossing in 1863, profited handsomely from the growing numbers of settlers, prospectors, merchants and military personnel who used this a shortcut across the wide, rolling valley. A trading post was also established there under the same name as the crossing. All during the 1860s, a military express ran through the valley, linking Fort Garland and Canyon City. When the Civil War ended the boom really began. The Homestead Act passed in 1862, along with the Indian Treaty a year later, opened up most of the valley.

Many veterans, some from nearby Fort Garland, took up claims, using the scrip they had been issued for pay. Many immigrants from Germany and other northern European countries who had served the Union cause were also entitled to this scrip. A few of the settlers were James Shultz, who settled close to Stewart's Crossing; Henry Backus, just northwest of present-day Alamosa; Peter Hansen, south of the town; Mark Biedel, on La Garita Creek; and a large group from the Company One Colorado volunteers, spread along Kerber Creek after 1865. Among the latter valley pioneers were Captain Charles Kerber, Lieutenant Walters, and George Neidhardt.

The valley was not yet self-sufficient, however, and pack trains from New Mexico provided food to the settlements for several years. The agreed-upon flour rations for the Indians usually turned out to be the moldy, wormy discards. Kit Carson complained from Fort Garland that a great deal of the flour from his home state of New Mexico was spoiled. This created a great demand for grain so, as historian Virginia

McConnell Simmons said, "The lush northern end of the San Luis Valley was quickly homesteaded and put to the plow."

Otto Mears took advantage of this, acquiring a contract to supply flour to the Utes because of his friendship with Chief Ouray. He sought out and found new markets for flour over this whole area, and built roads, toll roads and whatever else would help the valley and himself. This entrepreneur owned many stores and other enterprises and became one of the most prominent and wealthiest persons in the area.

The counties, formed mainly for reasons of political strategy or land production, consisted of Saguache in 1867; Rio Grande in 1874; Mineral County in 1893 (included here was Wagon Wheel Gap and Creede); Conejos and Costilla Counties in 1901; and the youngest, Alamosa County, in 1913. The "cottonwood" country now contains Adams State College within the valley's largest and hub city, Alamosa.

Historian Luther E. Bean, in his book *Land of the Blue Sky People*, summarizes the entire 4UR realm, including, of course, the mighty ranges of mountains, this way:

> The first colonists who came from New Mexico brought horses, cattle, sheep, goats, burros, swine, and chickens with them. They planted gardens as soon as possible, and made irrigation ditches from the streams to raise wheat, corn, peas, and beans. Chili was also grown extensively; no fruits were grown at first but they soon brought trees from New Mexico and started small orchards. They also had such wild fruit as currants, gooseberries, raspberries and chokecherries which gave variety to their diet. The livestock which these settlers brought with them was the same as found in New Mexico. Many of the horses were the ordinary bronco stock (Spanish mustangs), rather small but tough as steel. The cattle were mostly of the longhorn found in Mexico. The sheep were what is known in Colorado as Navaho sheep. They could rustle for themselves and live on limited pasture. Their wool was rather coarse. The burro, which has meant so much to the pioneers of the West (and indeed the world) served these colonists well. All of this livestock was quite well-adapted to the San Luis Valley.

Later, Midwesterners came, bringing huge draft horses to work wagons, timber and land with. It wasn't long until short-horn cattle

The bounty of the San Luis Valley. (Courtesy, Adams State College Library)

were introduced to improve the quality of beef. Alfalfa became an important crop, in spite of the short growing season, and is used to this day to feed valley cattle and even ship out surplus. It is said that the lambs grown on green alfalfa often top out the regional market. Potatoes, grown since the early pioneer times, became a very important food and money source, selling to the miners and saw millers in the nearby mountains and even beyond. Some deep irrigation wells were successfully dug and enhanced the production of this needed source of food for the entire region.

The wells also allowed the building of vast irrigation systems with attendant ditches and reservoirs. Large sawmills are now operating at Antonito, Monte Vista, South Fork and Fort Garland. The timber is cut in the surrounding mountains and hauled down to the mills. This cycle emphasizes the interdependence of the high and low areas— employment in the lower areas, timber from the high areas for houses and corrals, and on and on. Since the earliest days the milled timber has been returned to the mountains for the mines, mills and settlements there. An eternal circle of need and dependency—and its byproduct, love—has been created here, along with minerals, livestock, timber and agriculture, in the 4UR realm.

This realm includes the widely diversified San Luis Valley and the high surrounding mountains, many above ten or twelve thousand

Artesian well in the San Luis Valley. (Courtesy, Adams State College Library)

feet, with several topping fourteen thousand in both the San Juans and the Sangre de Cristos. Here is found a rare jewel of a region—of the world, a paradise of many sorts.

In spite of the earlier recorded historical people who crossed this valley, the spirit and cohesiveness of the entire area are somehow best represented by a single native, Jack Dempsey—also known hereabouts in his youth as "Kid Blackie." This Cherokee/Choctaw/Scots/Irish fighter was born in tiny Manassa, Colorado, in the southern part of the San Luis Valley on June 24, 1895. On that day, Manassa's principal resources of livestock, alfalfa, field peas, hay, oats and spring wheat were increased by a future heavyweight champion of the world, Jack Dempsey. He weighed in at eleven mighty pounds.

Although he was a peaceful child who loved birds, it was inevitable in those days that he get involved in altercations. Born to the battle, it would surely seem. One of his victims—at age seven—had to be revived by a veterinarian. There were many hardships for almost everyone in those days and about the only sport or pastime they had in Manassa was boxing. Jack didn't like it at first, but his older brother, Bernie, a fairly knowledgeable fighter, started training him. These early workouts soon paid off in the mining camps of the adjoining mountains when the family moved to Creede. His mother acquired a boarding house there.

Bernie knew fighting well but, because he had a glass jaw, couldn't make a career of the fight game. So he put all his ambition and talent into his brother, Jack. Bernie put Jack in a Creede boxing ring to take on all comers. Jack won all the fights. To help Jack strengthen his jaw, Bernie made him chew pine gum; to toughen his skin, Jack washed his face and soaked his hands in beef brine. Jack's muscles were toughened underground, mucking heavy ore into iron cars deep underground at Creede, Cripple Creek, Telluride and other mining camps of the time. He went on to win the heavyweight championship of the world from a giant of a man, Jess Willard. Jack weighed only 187 pounds when they fought on July 4, 1919. This 4UR-realm-toughened young man literally destroyed Willard and many thereafter.

He was truly a great representative of this land—proven possibly in just one incident. When asked who had hit him the hardest during his career, Jack said, "Gunner." (Gunboat Smith). When that fight was over and Jack became fully conscious on the dressing room table, Jack asked his trainer how long he had lasted. He was amazed when the trainer told him he had nearly beat Smith to death, winning in the second round. Gunner had hit him so hard in the first round that Jack didn't remember the rest of the fight! Speak of spirit: the land, the animals, the people all have it here.

If you wish to see and feel how the 4UR realm was long before man invaded it, first visit the Jack Dempsey Museum—it has to pay its own way—in the lovely little park on the main street of Manassa, then drive east over the Conejos River. Suddenly you will reach a spot of earth with no fences, no houses, no windmills, no anything manmade except the little paved road. Imagine the grass and other vegetation on each side of the road growing over the pavement and there it is: the feeling of newness, of vastness, of infinity with just the grassy hill in front and only the snow-capped Sangre de Cristos way out ahead and the malpai mesas all around. An eagle, a crow, or a hawk might fly as freely in the air as your heart, rare and precious. The diggings of the old turquoise mine are on the right and way ahead, in a low portion of the San Luis Valley, you can see the Rio Grande where it's so shallow that it forms marshes and a natural sanctuary for migrating birds. Ahead, barely visible, is the adobe Spanish village of San Luis that gives the valley its name. The spell is not broken yet—only changed.

Yes, the mother San Luis Valley and the father Rocky Mountains created a family of richness—if not always in hard cash, certainly in soothing the spirits of all who came and still come to fish, to hunt, to experience some of the West's first history and to be blessed by its greatest gift, the sorcery of simply seeing it.

CHAPTER 18

Natural Living and Some Tranquillity

At last, at last, Charles Leavell's moment of truth, the moment of his main dream's fulfillment, the moment of almost instant decision, was at hand. His heart drummed in his chest the same as it had twenty-six long years earlier at his first sight of the 4UR. Twenty-six years he had waited, dreaming his private dreams of a paradise found and owned. A place to rest, to fish and to have something rare for such a man: daydreams. Tranquillity. A sharing of the timbered mountains, the fresh flowing streams, the trout, the elk, and the spirit, the camaraderie, fellowship and soul kinship with those who returned year after year, and with his treasured family and the workers who also love it.

Earlier, this book described a few of those scattered places of the world, such as the Old Kingdom of Egypt, that are so special they become sacred. Do these places really exist? Ask anyone who has ever stood at the edge of Zambezi Falls on the border of Zimbabwe and Botswana, and heard the symphonic roar of this water-misted world, with rainbows arcing over it speaking of all the rocks and trees, and seen the streams of beauty of the world as they raise their eternal voice of unlimited magnificence. Then there is Machu Picchu, the "Lost City of the Incas," whose spectacular setting at eight thousand feet is probably the most lush, most amazing magic ruin in the world. The other surrounding peaks are covered in forest vegetation of so many blues and glowing greens, with such a richness of life, that they make

184

the viewer tread and speak softly so as not to disturb the mighty spirits that seem to be special caretakers there. One could stroll the beatific beaches of Bora Bora and do nothing but marvel at its radiance of sand, water and jungle growth. The 4UR has this same special beauty—a sacred feeling equal to any of the above.

Take this drive from the 4UR headquarters: Go east across the Rio Grande onto highway 149, turn north toward Creede for about a mile and a half, then take the Rio Grande National Forest road to the right. Climb through and around mighty rock bluffs, lush mountain meadows, pine trees and groves of aspens with their lower branches grazed bare by elk, and it will become clear why the 4UR should be ranked as one of these sacred spots. Turn off the well-graded forest road onto Pool Table Mountain and at just the right spot get out of the vehicle and stand on the edge of eternity, looking way down and west to the valley of Goose Creek.

Early in the month of June, it seems like a cliché has come true: you feel as if you can see forever. The vastness of the drainage system that makes up the 4UR and Goose Creek land will slowly envelop the beholder. There are great packs of high snow melting and running down the bottom of timbered canyons so large that entire good-sized towns could be lost in them. The high country meadows, as green as giant emeralds, are spotted throughout like God's golf course. It's almost more than one's breath and eyes can encompass.

Also in plain view is the great Continental Divide. That rocky backbone of our nation is clearly marked in an amazing abundance of still-frozen snowpack like a mighty white-backed snake made of stone, crawling its ponderous way all through the United States on a curving route of preeminent power and allure; motionless, yet forever moving to Canada and even on through that nation. As the sun creates rays of golden, moist light through the scattered clouds that caress and reveal the soft curves, the hard bluffs, the bounty and the blessedness of this colossal, divine construction, one gets the goose bumps of Goose Creek and the 4UR. No wonder everything else is forgotten. It feels like witnessing the entire world just an instant after its creation.

On a special day in August of 1970, Charles and Shirley were taking a short vacation at the 4UR, escaping from the stress of everyday business in El Paso. The ranch didn't belong to them yet; they were guests. Before they had arrived, Charles had been thinking about breaking up his holdings in favor of his children and grandchildren (some yet to come). In fact, he had already started the process to a small degree. But now he stood in his favorite stretch of Goose Creek with only one thought on his mind—trout.

The creek caught the light in millions of little diamond suns that sparkled an instant before disappearing and reappearing in a rhythm of light and shadow that both soothed and exalted at the same time. Charles had long ago learned to listen to the endless voices of moving water. He could read the water as he did a book. He recognized the curious line of creek water that spelled out a feeding channel. He had just flicked the fly into it when a rainbow rose from underneath and took it. Charles stood his ground, gave the line the exact right jerk and felt the weight of the fish well-hooked and fighting with the valor for which they were created. It broke to the surface several times before Charles landed it with his net. It had been a hard fight but the weight on that line was more precious than striking gold to Charles. "Ah, heaven is in a good fishing stream," he muttered to himself. His heart pounded with respect as he carefully released the trout back into its own shimmering haven to thrill another angler another day.

That night, after dinner in the main lodge, Shirley retired to their cabin and Charles visited with Chuck Davlin, business manager for the 4UR's owner, Allan Phipps of Denver. Charles told Davlin that if Phipps ever decided to sell the ranch, he sure wanted first crack. Davlin said, "You got it."

Charles said, "I'd about given up that I'd ever get a chance to buy it, but a strong part of the dream remained no matter what. All my life I'd wanted that one special trout stream, and my own private place to get away, fish, look at the mountains and meditate. Ever since 1944, no matter how I'd looked and tried other places, this was the one for me. I'll admit that, as sinful as it may sound, I coveted the 4UR and Goose Creek."

Anticipation.

The very next morning Davlin called Charles aside and said it was hard even for him to believe, but Mr. Phipps was ready to sell. He gave Charles a price at the same time, saying that Mr. Phipps didn't intend to negotiate at all; this was the final price. Here it was. A decision had to be made. As much as he wanted the ranch—to the point of toothache pain—Charles still dealt by saying he'd write a check for $10,000 for a thirty-day option at Allan Phipps's price. Davlin called Phipps in Denver and he said okay.

The place was really run down, because Phipps had taken more interest in another place across the Rio Grande, La Garita, which he owned and where he had formed a private club of his friends from Denver and elsewhere. Charles knew he was going to put up the rest of the money, but wanted to have a brief time to compute the repairs and analyze where he really wanted to go with the place personally and how much it would cost to make it first-rate. He could go with nothing less.

As they drove away, back toward El Paso, he asked Shirley how she'd like to own the place. He had mentioned it so many times that she didn't pay much attention. Then he said, "Well, I just bought it."

Then she paid attention, saying, "I really didn't expect to get a fixer-upper."

Charles took a quick glance before they crossed the bridge over the Rio Grande and was happy to see the blue eyes of his most loved one smiling. The drive home to El Paso was several feet above the pavement in pure mountain air after that.

Although he had never felt better in his life, before he could get all the funds and other necessaries ready for the much more than a million-dollar permanent purchase, Charles was knocked flat with a congested lung where a blood clot had formed. It was very serious and the doctors said he could die any second. He says simply, "I thought I was a goner." Coming from Charles, the one who never put up with defeat, this was indeed serious stuff.

In spite of the doctor's protestations, Charles managed a call to his friend, Willy Farah, an El Paso manufacturer of pants. Farah was in California, so finding him became a terrible risk. Charles's love for the 4UR drove him to take that risk, and finally they found Farah.

Charles said, "Willy, you've got to listen to me, I've got an option on a ranch and you don't have time to go check it out. Take my word for it and put up fifty percent with me and let's buy it. I'm a sick man here in the hospital and I need a partner."

What's amazing to most people is common to Charles H. Leavell. Farah said, "You got a deal." Charles lay back on his pillows knowing Farah had the knowledge and friendship to handle it. He did, even though he didn't know exactly what he was buying. Charles could now put what little energy he had left into stalling off death. and he did. It was another miraculous recovery, just as he had done so many times in business dealings and dire physical situations throughout the years.

Detailing all the trading and negotiations that took place is unnecessary. Farah knew little about ranching and soon sold his 50 percent to Darcy and Ruth Brown. All the 4UR land was given to the Leavell children and the Brown children, who jointly entered into a seventy-five-year irrevocable lease with the 4UR Corporation owned by the Leavell family. The Browns still own the adjoining Humphrey Lake property.

Finally, Charles had the four thousand acres along the eight-mile creek. Except for his good friends to the west, they were surrounded and protected by hundreds of thousands of acres of the Rio Grande National Forest and wilderness area. As Charles so succinctly put it, "At last we found and possessed our paradise-yet-to-be-improved." The truth is, Charles had always thought of it in the plural, for his family and the patrons who would also become like family—some of whom were as special as the place itself.

CHAPTER 19

Repairing Paradise Just a Little

 Everyone through history seemed to know of the specialness of the 4UR and its environs. Members of the Hayden Survey party at Wagon Wheel Gap described the Utes' attitude as follows:

> Indians used the southern hill as a point of lookout. A long wall runs along its entire northern edge, and round towers two to five feet high are placed along it at different intervals. Walls on the south afforded protection against anyone advancing up the gentle slope from that direction. The fortifications are well conceived, and the locality chosen with judgement. Numerous fragments of chalcedony and jasper furnished material for the manufacture of arrows and spearheads, a number of which were found in the old stronghold. Mr. Wilson found one arrowhead of obsidian, which certainly was never obtained from any locality in Colorado, but must have come from New Mexico, or even further south.

Most of this fortification is still intact and shows the huge amount of labor the Utes went to, besides several bloody battles, to protect what is now the 4UR.

In 1893, the *Creede Candle* reported that the hotel at the Hot Springs was attracting Creede people all days of the week and on Sundays the pilgrimage "amounts almost to an exodus." In 1900, a letter to the editor of the *Denver Times* urged, "Away with your Newport! Away with your Manitou! Away with your ocean breezes and delicious salt air! Give us some Wagon Wheel Gap on toast!"

The touting brought in Charles G. Dawes—later vice president of the United States under Calvin Coolidge—for a fishing stay in 1905. Ten full days. It was said to be the longest period of time Dawes had ever spent on vacation at any one place. U.S. senators Phipps and Humphreys also found the mountain resort a magnet.

In 1904, the St. Louis International Exposition awarded a silver medal to Wagon Wheel Gap Springs. By 1908 a casino was available for guests. It had a large dance floor, fireplaces, a piano and billiard tables. A brochure of the time assured the public: "No venomous serpents of any kinds are found in this region." A masseuse was available and daily mail, telephone and telegraph were touted at the hotel, as were a nine-hole putting course, tennis, dancing, mountain climbing and, of course, the soaking in the springs.

In August 1948, President Eisenhower brought his entire immediate family for a relaxing stay. Noted fishing instructor and guide, and long-time 4UR guest Bill Geis said that King Leopold of Belgium had visited there in the early days. Yes, the 4UR surely had a prominent past, but it was the present and future—especially the future—that Charles Leavell was to concentrate on with all his being.

First of all, the buildings, including cottages, corrals and almost all the other manmade structures, were run-down. This meant that Charles and Shirley had to do large and careful planning to get the atmosphere and comfort combined the way they wanted. They wished to keep the old feeling but with new neatness and comfort. They installed complete new water lines and disposal systems. Then there was the repair of riparian areas on the banks of Goose Creek and the ultimate construction of water-slowing ponds that would improve the habitat for fish and consequently the excited pleasures of addicted and new anglers. It was, as Shirley had half-kidded Charles, a real "fixer-upper" job. They were, of course, up to it.

Fortunately, along with the property they inherited a cattleman by the name of Ed Wintz. He was a top-notch cowboy and horse trainer, but had the knack that comes from the long years of necessity that almost all working cowboys have—he could repair anything. He was also a great hunting guide and guest wrangler into the bargain. In other, more pointed words, he was priceless and essential to the Leavell's 4UR plans.

Ed Wintz and his wife, Dea Jean, spent, all told, forty-six years on the 4UR before retiring to their own Lonesome Bear Creek Ranch above Creede, where he still outfits and guides. To this day, in the Creede country any horse Ed has bred or trained is known simply as "a Wintz horse" and is highly prized.

As we've already seen, one of Charles's great strengths was his ability to acquire the highest class help and maintain a two-way respect and dedication to any project. Because the 4UR was his dream endeavor, those who were with him became more than fruitful and skilled workers; they became family.

There is another couple who gave not only gave their skill and love to the 4UR, but their souls as well: Joe and Katus Walton. How this couple got together, and then arrived to serve and be served by the 4UR's special lands, is another strong strand in the interlacing of this powerful web of destiny.

Beautiful, blonde, twenty-year-old Katus had been torn from her family in Hungary to work in a hospital in Neurburg, Germany, as a nurse during World War II. It was a horrifying time, but finally the American liberating forces approached. The SS troops fought hard and for two days the shells screamed back and forth above them. The civilians had no idea whose prisoners they would become but prayed it would be the Americans.

When the liberation came, Katus was making a medical delivery on a tray. Shaking with fear when she heard a loud voice shout, "Halt!," she almost dropped the glass containers. All her horrified vision saw was the "enemy" and his tommy gun. As she recoiled, a gentle hand touched her blonde hair and said, "It's okay. It's okay." The GI gave her some candy bars and chewing gum for the Hungarian hospital staff. For them, the war was over. As Katus later explained, "The gypsy played his violin. From the first meeting with the gentle GI, I was determined to get to the United States."

After she had taken a nursing position with the 116th Station Hospital, Katus began dating Colonel Joe Walton. He was in charge of the military police force and was deeply in love with Katus. Knowing that she returned his emotions, he said, "I know a way to solve the problem of you getting to go to America. Marry me." And so they did, and a great marriage it was.

When Charles Leavell became aware of the couple, Katus was catering manager of the famed El Paso del Norte Hotel—a prime position—and her husband, Joe Walton, was general manager of the elite White House establishment in the border city. They were hired by Mr. Leavell to jointly manage his beloved 4UR. A powerful compliment, indeed.

In spite of the initial repairs done by Sharp, the hundred-year-old buildings had begun to run down. It was just a fishing camp in this beautiful spot until everyone slaved and sweated over Leavell's vast improvements.

The Waltons were loved and beloved way beyond just being managers. Joe had a great sense of humor, along with being extremely handsome. He was an asset of entertainment and softly tough as a confidence builder for all the guests, from toddlers to ancients. Katus did everything humanly possible, from treating and consoling a child with a skinned knee to fixing special picnic lunches for all ages. She missed nothing that would enhance a guest's stay at the family 4UR.

Through the years, the Leavells and all the hands and guests depended on Katus's gracious giving to such an extent that when she and Joe retired to return to El Paso, more than half the guests swore they could never return. It would be too sad with them gone. After all, they had lived, loved, and given all they had to the people, animals and the land for seventeen years. It was a tough job, but the Leavell family— along with the fine work of the replacement team, Kristen and Rock Swenson—has brought those faithful guests back into the ranch family.

Joe died a year after retirement and Katus became the postmistress at Santa Teresa, New Mexico, near El Paso on the Mexican border. That town is now being developed as a major border crossing between the two nations.

Even now, Katus's deep love for the 4UR cannot be expressed without tears dampening her wise and lovely eyes. She recalls each Sunday during the guest season as being both so traumatic and so joyful that she can barely express it. That was the day when the great 4UR mission bell rang for those departing. Katus knew that some of the oldest, who were like kin, would never return. And she wondered if she would ever again see the children she had watched and helped grow up.

Then, in the spring, the returnees and Katus watched anxiously for the appearance of others, wondering how much the youngsters had grown, how the health of the elderly had held up and what was behind the general expressions of joy or distress on all those in between. She cared so much about the ranch's "family" that every week her emotions were dropped and lifted to an extreme degree. She says, "After all these years, I still listen for the mission bell every Sunday morning."

Katus understood the reality of beauty and dignity at the 4UR and she also understood and personally saw and felt the great spiritual presence there. Her being is part of Goose Creek, the mighty forests, the earth, the buildings and all who came on her watch. It will remain for those yet to come.

So, slowly, surely, with great love and great effort, the ranch became knit solid and a haven of security for all. The cabins were carpentered back to new. Some new buildings were added but the rustic look was kept. The grounds and the main lodge for dining, drinking, dancing, visiting and pure enjoyment were put in shape and decorated with fine original oil paintings—dominated by those of the Leavells' close friend, Tom Lea. The corrals, the tack rooms, the stables and the corrals were all fixed up sturdy, neat and very usable, without any look or feel of a contracted, brand-new facility. The old feeling had been maintained here with skilled artisanship just as carefully as in the living and dining quarters. Homes for the managers were also built or put in proper order. The circling roads of the compound of pleasure were also carefully graded or paved. The horse pasture fences and gates were evaluated for grazing quality, fencing and gate repairs so that all who resided here (including the saddle and pack horses) would have the same feeling of care. The springs that had fallen prey to the constant wear and neglect had to be repaired. In some cases underground pipes were laid to pump the hot water to more useable spots.

Of course, Charles's first and final love of this land was and is Goose Creek, on which all the other things depend. Through the years, with dedicated help from many but especially from his son, Pete, the large, talking, sparkling creek has become a family member as well. The Leavells knew that if it were developed and maintained for the maximum output of wily trout and effervescent beauty, all else would be profitably and environmentally enhanced. The horses, the elk, the bear, the deer and the marmots would all drink and feed on its productive banks and adjoining lush meadows, in harmony with all the songs of wind, water and their spirits, to the benefit of everyone and everything.

All this came to pass—slowly, sometimes painfully, most often joyfully. Charles Leavell was at the peak of his financial expertise in the 1970s. He took actual millions of dollars and many, many millions of dollars' worth of time and applied them toward his view of the resurrection and near-perfection of the 4UR. Even as he divided up his properties among his children and grandchildren, he spent heavily on enhancing the best, world-class fishing areas of Goose Creek alone.

From the beginning he had adored, and absolutely respected, his wife. This not only proved he was a wise and loyal man, but also enhanced their sharing of all the business sacrifices and now the great dream in this precious, chosen valley. With his love so strong for Shirley, Charles could never have entertained the thought of taking a mistress. Nevertheless, he had acquired one and he gave his own kind of unique love to her just as he always had to Shirley: her name was Goose Creek. She was an expensive mistress but indeed returned more than she received to all who set foot or hooves or claws here. Like the Leavells, who have given so much to the world, the creek has through its ancient rhythms and actions, given even more.

The elevation at ranch headquarters is eighty-five hundred feet. This ensures a cool season that runs from June 1st to September 26th. At first the Leavells ran cattle under the care of Ed Wintz, but after a few years they cut them out for better use of the land as a guest ranch.

Cowboy-outfitter Ed Wintz has connections about as close as one can get to the ranch. His father had lived five miles south up Leopard Creek (a fine feeder creek that drains into Goose Creek) that is now noted for its native cutthroat trout. Ed's father attended a little red schoolhouse near the 4UR, and so did Ed Wintz for a short time. Ed remembers well the story his dad tells, for if the outcome had been a little different Ed would not have been born to live and fish on Goose Creek. It was Christmastime and Ed's father and Ed's aunt were attending a holiday party at the tiny schoolhouse. A snowstorm hit full

Shirley and Charles Leavell, 1990.

Longhorn pets on the 4UR.

blast. When they got ready to leave the happy gathering, they found that their sleigh horse had pulled back, gotten tangled in the reins and rope and choked to death. So the two of them had to walk home the five miles up Leopard Creek. They came very close to freezing to death.

During the time his family lived on Leopard Creek, Ed's granddad made a living cutting and hewing railroad ties. Ed's father said grandpa could cut and finish twenty ties a day, for which he got $1.00 apiece. Ed said he'd hate to think he had to make twenty ties a day even using a modern chainsaw. Tough genes in the Wintz family for sure. Around 1920, Ed's grandpa moved the family to a cabin on Goose Creek a mile and a half above the 4UR headquarters. That cabin is now known as the "flimsy stocking." While there, grandpa used his six-horse team to haul cement to Humphreys for construction of the dam, which was completed in 1923.

Ed went to work for Arthur Sharp from Colorado Springs in 1953. Even then the Wagon Wheel Ranch, as it was called, was a combination cattle, guest and hunting ranch. The Sharps owned the ranch back when the Leavells first visited.

"Arthur Sharp was just like his last name," Charles said, "and his wife, Geneva, was just as dedicated to the ranch."

Sharp built the modern ranch, first tearing down the Palmer Hotel and then replacing it with the current fine lodge. Geneva did all the home-style cooking and a hundred other things. Ed Wintz was soon made foreman and continued on with Charles Leavell in the same valuable capacity.

Even so, these things were not Ed's main interest. As early as he can remember he liked to ride horses, wild or tame, and anything else that was big enough and had four legs. When he was around eight years old, accompanying his father running a trap line, he had an unintentional ride. They came upon a yearling doe hung up in a fence with both hind feet. They were facing uphill when his father lifted and freed the deer. Instead of going on uphill, as they both expected, the deer did an about-face, ran under the loose bottom wires and right between Ed's legs, then raced downhill with Ed Wintz riding her backward. Ed said, "The ride only lasted about fifty feet but it was the fastest fifty feet I ever flew on a four-legged animal."

Besides all those fine-bred horses he rode and trained, Ed Wintz is one of the few people in the world who can honestly claim to have ridden not only a deer, but an elk as well. Ed was helping a game warden, Punk Cochran, trap elk for tagging years back on the 4UR. One cow elk that had been trapped refused to jump out into the net for easy handling. So both men, being cowboys, headed and heeled her with their catch ropes and stretched her out on the ground to apply a tag to her ear. Then they skidded her out of the trap to release her. All seemed to be going well, except that Ed Wintz suddenly had an uncontrollable urge to ride an elk, so he jumped astride just as she rose. The elk naturally took a downhill run in the snow, bucking or "crow hopping," as it's called when it's not round-corral or rodeo-type action. In other words, easier, not so high and not so crooked. It wasn't as pleasant as Ed had envisioned. He grabbed at the elk's skin, trying to stay on the animal as they headed downhill, gathering speed with each hop.

Ed was really having fun until he saw a barbed-wire fence rapidly coming up. Fortunately, the elk's loose skin did him in. He lost his hold and hit the snow. He said, "I rolled into a snowball for an easy landing. The elk, rid of the human fly, turned away at the fence, saving herself as well."

When the laughing game warden watched the snow-caked cowboy walk up unhurt, he said, "I didn't know you planned to do that."

Ed said, "How could you? I didn't know it myself."

Ed, now in his late sixties says, "One more time, I'd like to ride another wild animal, the moose. If I can talk my pretty wife into helping my aging body up on one! I figure this would be a better way to check out than going out to sit on the ice like the Eskimos do."

Every cowboy makes a fool out of himself sometime during each working day, but some things are so common they are forgotten, especially if he's by himself. But there's one that Ed remembers, from the forest high country adjoining the 4UR. A rancher friend had a wild cow they just couldn't gather, so Ed—always trying to do good for friends—decided to corral her that winter.

He used up a couple of good horses and all his cow sense, but he finally got her penned. He proudly called the rancher and his visiting brother volunteered to help load her up. Now, as already related, this was a wild cow. Ed stepped into the corral first and without any other thought but having fun decided to play bullfighter. The cow was ready; she instantly put her head down and charged Ed from the far end of the corral. She had time to reach great and powerful speed. Ed stood his ground, sidestepping her as she closed her eyes to butt him into oblivion. The brand new bullfighter yelled, "Olé!" as she whizzed by. She almost instantly reversed direction and charged him again. Ed barely sidestepped her this time and his second "olé" was a bit weaker.

He said, "On the third pass she had me figured out. She came churning the corral dust, right at me, with even more speed, head held high and with her eyes wide open. She hit me dead center in the chest and I went flying through the air without any ease and no breathe left in me. Naturally the rancher friend and my brother were hanging on the top rail of the corral with uncontrollable laughter. Gasping for any kind of air, I made it to the fence before she could charge again."

When his two companions could speak, they asked Ed if he'd do it just one more time. He could not answer either way. He was trying to suck some oxygen into his lungs and could not accommodate them.

There never lived a cowboy who didn't have some kind of experience with a coyote that was indelible in his mind. Ed was riding out with a favorite cow dog to gather a few head when he heard his dog growl. A coyote that was following them kept moving in just close enough to the dog to agitate, but not endanger himself. Ed rode along, looking back and wondering what the outcome of this strange encounter would be. He was just as curious as the coyote, but didn't want anything to happen to his partner, the dog. But it did: The dog could no longer take the teasing and charged after the coyote at full speed, no doubt so enraged by now that he thought the coyote would be just so much limp hide when he finished chewing on him. Obviously the dog hadn't heard about how the wily coyote lures dogs into a multiple ambush and has his buddies do the dirty work.

The coyote must have read the dog's mind and expertly ran right under the bottom wire of a barbed-wire fence. Ed looked on in fear as his dog neared the deadly cutting fence at full speed. The cowboy was sure he would be cut in half or crippled for life. He held his breath as the dog hit the fence at full speed, but was somehow cast sideways by the impact and went through the fence sideways and uncut. However, the impact had taken the chase out of him. Ed watched the coyote run up on a little hill and then stop and look back a moment. His dog was safe and well and in few minutes would be ready to help him gather cattle. Interesting and lucky day for all.

One year, after the 4UR guest season was over, Ed went on a bear hunt with a bow. He, his family and his brother were staying in their home in the mountains above Creede. He'd had no luck so far in even seeing one until just at daylight. Ed and the entire household were still soundly sleeping when Ed heard his German shepherd barking frantically. He got up, put on his house slippers and, clad in pajamas, went to investigate. It was still too dark to see clearly, and he was only half awake, but an experienced outfitter like Ed Wintz knew the growls he heard were not the dog's alone.

A huge black bear was eating the rest of the dog's food from the bowl. The dog, being no fool, was simply running around and around the mighty bear, making only vocal protestations.

Ed picked up his bow, fitting an arrow into it, and walked out to bring an end to his bear hunt. The bear raised his head before Ed could hardly get off the porch and with a "whoof" charged. Ed retreated into the house, barely escaping the bear and only having time to slam the screen door behind him. He quickly set his arrow, realizing that a screen meant to exclude flies would do little to stop the bear, who already had his nose up against the thin wire between them. The bear started to raise up to smash the flimsy door when Ed let loose an arrow that struck the bear right in the jugular. The beast turned and ran to the nearest tree, climbing it. The sensible dog now became much braver and began jumping up the tree trunk as if he wished to pursue and conquer.

Ed, having more experience than the dog and not being a bluffer himself, sky-lighted the bear and placed two more arrows to ensure everyone's safety. The bear fell from the tree dead. When Ed and his brother skinned the bear out, they found a lead bullet embedded in his rump that was causing infection. That, of course, had slowed down the poor creature's hunting ability and explained why he was seeking any easy food he could find.

They measured the skull. It held the world record for the next two years in bow hunting. Later, when someone asked Ed why he didn't use a gun, which would have been much surer than the bow, he forthrightly explained, "I didn't want to wake up the kids."

Besides undertaking all these adventures—and many more—Ed Wintz broke and trained horses, wrangled the dudes, guided the hunting and did almost anything and everything else needed to help make the 4UR what it is today. So it is little wonder, then, that Charles Leavell recently stated, "Ed Wintz was a very great part of the 4UR history." Ed himself turned most tragedies into fun, so one can imagine how much joy he gave to the guests over the years.

The Leavells stopped raising cattle at the 4UR to assure lots more grazing for the forty guest horses. They also figured it would increase the wildlife, which would be more enjoyable for everyone. Ed Wintz has never commented on how he felt about the change. However, if he wished he could probably add a few fiery notes about ending the hunting parties.

A group of the wealthiest, high-ranking politicians from a south-of-the-border country had called and made reservations for a two-week hunting vacation with their families. These "families" consisted of a busload of girlfriends and mistresses. The guests took control of the ranch. They partied day and night and with some arrogance demanded that Ed and the other hands drive the wild game right by them. Ed and the hands politely declined, and in fact there was very little hunting for four-legged wild game during the drinking, dancing and romancing. The accompanying noise was of such volume that one hand said all the sensible wilderness creatures had probably migrated to Wyoming. Those "guests" would have turned the place into a cockroach den if they had stayed another two weeks, according to another worker. They were forbidden ever to return.

Charles Leavell had been thinking of closing down the hunting anyway, but had kept it to himself because Ed and others loved it and were so expert at outfitting and guiding. However, he now had a legitimate reason to do so, even though it cut a couple of months off their profit season. They could now concentrate entirely on the four unique summer months, giving total attention to the summer guests. In the long run, his decision turned out to be right. One can hardly imagine a place of this sort where the guests feel more at home or enjoy more bountiful gifts than those offered at the 4UR.

Charles Leavell also made the decision never to advertise, although they do have a brochure to hand out to those who ask. Another mainstay at the ranch was, and certainly still is, Bill Geis. Bill is known around a lot of the globe as a world-class trout fisherman and instructor. He has partially retired now and has a fine log house right on the banks of the Rio Grande at the nearby town of South Fork, but he's at the 4UR several days a week, giving lessons in casting and all-around fishing. He also has fly-tying sessions a couple of nights a week in the main lodge.

Bill is in his early sixties but is lean, quick-moving and enthusiastic about the 4UR and the Creede area. Professional outfitters from New Zealand, Australia and other environs have paid the ultimate compliment by traveling all the way to Colorado to study with him. His obvious joy and knowledge of his artful craft make it a memorable experience just to be around when he is working at playing (or vice versa). He probably knows the area's mines, hunting camps and best fishing holes as well or better than anyone alive. He was also a dude wrangler at another ranch across the river years before Mr. Leavell acquired the 4UR.

It is easy to see why Bill Geis became such an integral part of the 4UR. His spirit and that of his charming wife, Betty, just fit perfectly. Everyone loves and respects them.

When he gets started telling Creede and 4UR history, his enthusiasm is absolutely boundless, but he becomes serious on occasion. "You know the white man came in here early on and destroyed lots of the country. They cut down all the timber for stoves, houses, mine timbers and railroad ties. Then they dynamited the fish to feed the miners—sometimes five or six thousand pounds a day out of the Rio Grande here alone. But now it's back somewhat, thanks to a few people like Mr. Leavell. I wish they'd get all this history down now, because it's going fast."

He grows slightly more somber when he says, "Most people don't like history. All they're concerned with is the TV, buying a new car and having a lot of money." Then he'll sparkle all over again and, if you wish, tell you where you can get the big one, take you there for a modest price (he does outfitting on his own) and show you how to catch that prize, elusive, slippery pan-filler on any stream in most of the 4UR realm and all of the Creede area. Bill Geis was a prize catch for Charles H. Leavell and remains so for all who use his knowledge and his love of mountains and streams.

CHAPTER 20

Time of Rebirth
and Friendships Rare

The bathhouse at the 4UR was built in 1904 and is reputed to be the oldest concrete structure in Colorado. In the summer of 1995, the Leavell family had a meeting to discuss its future. Fortunately, they voted to restore the historic building. The future of the abandoned fluorspar mine, with its picturesque mill, is still in debate. After restoring all guest facilities to modern standards, yet retaining the atmosphere of the adventure-filled years that brought the ranch through its noted previous owners right up to the Leavell family, one thought was dominant in Charles's mind: he truly wanted everyone else there to share this special pleasure. Vicki Goldberg, writing in Conde Nast's *Traveler* magazine, said of the Wagon Wheel Gap days, "The town was famous for its mineral springs: the spa took care of your health and good trout fishing took care of your soul."

Those are pretty much the Leavells' feelings today. Ms. Goldberg further enhanced Charles's view by writing, "After all, Izaak Walton once pointed out that Jesus chose four fisherman for disciples."

It is widely known among the world's fly fishers that the feeling of ecstasy gained while watching gurgling waters soothe the eyes and caress the legs has healing powers that would make the greatest masseuse jealous. Then there is the anticipation of the water's creatures that could only have been designed by a divinity—waiting, teasing, then striking in a sudden wondrous sensation of perfected fury. No matter how many times the fisherman experiences it, each strike on that artificial fly is a connection, complete in that second, to the primeval past that courses through the veins, nerves and genes in an instant recall of ancient survival skills and sport. To fly fishing addicts, it is unequaled by anything else. No wonder it is called a sport nigh onto religion by so many. One touches eternity just as the swift and glorious trout touches the fly. As an example, Robert O. Anderson, former owner of the famous Ladder Ranch near Hillsboro in southern New Mexico (now owned by Ted Turner and Jane Fonda as a two-thousand-head buffalo ranch) and former president and major stockholder of Atlantic Richfield Oil Company, comes every year for two weeks to the 4UR with his family and twenty-seven grandchildren. Charles chuckles, "It takes the entire ranch to take care of this bunch, but they sure enjoy it, and so do we. Robert O. is used to being the boss so he's always ordering people around. He's king of the hill when he's at the 4UR. R. O's grandson, Jay Smith, and Charles's granddaughter, Jennifer—Pete's daughter— are married to each other. Both families are very happy with this union.

Bill Geis, the fishing instructor and guide, tells of another experience with the world-famous opera singer, Lawrence Melchior, who once came to both hunt and fish.

"Mr. Melchior was the fattest man I ever saw. Of course, I hear you have to have lots of weight to blast out a song like he did. Anyway, he couldn't handle the walking and riding to properly hunt big game. So we just set him in a chair under a tree with his gun and told him we'd try to drive some game by. Of course, you can't drive these Rocky mountain deer and elk. They make their own trails. After about three days of waiting he gave up and took to the creek."

To the surprise of many, the singer turned out to be a pretty fair fisherman, landing a couple of eighteen-inch rainbows and suddenly enjoying himself immensely. That's the *musical* power of fine trout water.

Anyone who listens to Charles Leavell talk about any portion of the 4UR is, of course, infected with fishing fever, but there are other things he says about it. "I've ridden it, walked it, hunted it and fished it. I've studied the Indian fort and tent rings, I've gone all through the mine and mill. I bought the mine from the C F & I and we've got to find the rest of the historical papers on it. You see, historically it's a great old mine. It's got big old cave-ins at the top of the thing. They're hidden from the view of the guests. It looks like a small atomic bomb hit the earth up there. It is really strange to show the mine and mill to people. They love it. You can't believe it. Oh, I love it all so."

Charles found a man who knew how to improve a stream with great skill. They built big overhanging banks, shoring the dirt with heavily braced pine lumber. (Pine doesn't deteriorate much as long as it's underwater.) They covered these over with rich soil, grass, wildflowers, aspen thickets and everything else indigenous to the valley. After awhile all these root systems intermingled and the improvements became permanent and an undetectable part of the landscape.

Charles says proudly, "You can't tell they were ever touched by human hands. But underneath is a cave with big fish feeding. And boy do they come roaring out of there after your fly. It is fun!"

They also learned how to move large rocks—chair-size and even bigger—so they could form more dammed-up ponds. They are spaced to take care of all the numbered fishing locations. The cost in dollars was large, but the benefit to the main artery of the 4UR valley and all its inhabitants is priceless.

Charles recently had a friend from the highest court in the land, the Honorable Byron White of the United States Supreme Court, down for a few days of fishing. White's obvious approval was a pleasurable seal on all the labor and planning they had done over the years. (Justice White has since retired from the Supreme Court. The courthouse in Denver was named after him. He not only made the highest court in the land, but also left his mark on some who still remember him as the football star called Whizzer White.)

The 4UR has fifteen, one-half-mile fishing stations that the guests who want to fish draw for each morning. They change locations after lunch. This gives everyone a fair shake at the best of the waters, although the worst are still very good—just heavier timbered and harder to fish. The half-mile stations give a feeling of freedom and privacy while indulging in these wondrous waters. Above stations number eight and number nine, the stream boxes up and all the trout are native-breeding. On these upper reaches, a lot of the stream is white water, as it comes down out of the mountains at about a 4 percent grade. The

A 4UR guest and his catch.

water is really in a hurry here to get on to its mother Rio Grande. But the big wild fish are there. All that's needed is desire and stamina. The lower half-mile stations are stocked with about a thousand pounds of fish a year.

Charles says, with obvious joy emanating from those steady blue eyes, "It's in magnificent shape. Magnificent." He explains, "You see, the fish, the cold fish that survive the winters, are wild the next season. The first year they're fairly docile. But we have to stock it. We have fifty guests with only twelve weeks time, we're about 95 percent occupied, and out of the fifty guests I'd say at least two-thirds are fishermen. That's five hundred in a summer. So, there is no apology for the stocking of the lower levels. There are uncounted big wild fish to catch in the lower nine half-mile stations. In July and August we're so full we shake the dice for positions. Everyone likes it that way. No favorites, you see. We have fifteen stations in all on the creek and four on the Rio Grande. A man and his wife will fish the same beat. There is nothing more beautiful than to see a couple completely engrossed in the stream.

Charles Leavell doing what he loves best (1991).

"The lines flicking back and forth, with the sun creating little thin auras of moving golden light in the mountain sun, is a form of magic, all right. And then the flick of the line at the strike, the play of the line, the rainbows blasting a frothy hole in the surface or a brown plunging for the darker depths, as the watching spouse, silently, and often verbally, cheers the other on. A rare form of sharing, of bonding together with nature and the wilderness from which we all came so long ago."

Charles Leavell has always known that success comes only from sharing. His enjoyment of the pleasure of others, here in this fairy-tale land of beguilement, is unlimited and enormously fulfilling to this man, who is just as special as the 4UR itself.

Tom Lea first pointed out the 4UR to Charles back in 1944, saying he had heard it was "great fishing." One cannot fully absorb this unique friendship without knowing of Tom Lea and his vast accomplishments. First, the Leavells and the Leas have been best

The reason for Mr. Leavell's success is evident in this photograph.

Charles Leavell and Tom Lea at the JM. Big country, good cattle, good company.

friends for over fifty years. That alone is a strong statement of faith. Even so, it would be almost impossible to overstate the enormity of Tom Lea's gifts to the world.

Lea, like his friend Charles, was born in El Paso and deeply loves both sides of the border. This often-harsh land has produced some of the greatest and most original artists of our country. There is Louis Jimenez, the sculptor in his own invented form of brilliant plastic; and José Cisneros, the illustrator and painter of classic horses, vaqueros and the searing landscapes of the area. Hugh Cabot—of the Boston Cabots—came to Terlingua on the border and painted and drew hundreds of fine portraits of the last of the living vaqueros; then he moved on to Tubac, near the border in Arizona, to continue his portrayal of the powerful landscapes and border people there. Dr. C. L. (Doc) Sonnichsen, who wrote and taught at UTEP of the border's history and beyond, mentored students and proteges such as Leon Metz, Dale Walker, Brian Wooley, W. C. Jameson and Nancy

Hamilton, who carry his traditions of border history proudly on into the next generation. Carmac McCarthy moved there to create his greatest work, *All the Pretty Horses*. John Houser, the noted sculptor, and Tom Lea's late friend, Carl Herzog, the world-renowned bookmaker of El Paso, also inhabit this strong list of border artists. There is Eugene Cunningham, who wrote the long-selling book *Triggernometry*; and Gene Roddenberry, the creator of *Star Trek* and its many spin-offs covering the world in film like no other known artistic endeavor. There are too many to name each individually. Still, one need not hesitate to state that as a combination writer/painter/illustrator, it would be difficult to find anyone in American art who could possibly surpass the amalgamated skills and deep emotions of Tom Lea's creations of both visual and verbal masterpieces.

Tom Lea was born July 11, 1907. "The eleventh day of the seventh month," Tom says, "I should be lucky as hell at craps." Well, he has been, with the greatest crapshoot of all: painting and writing for a living. Not only that, he is one of the few who ever lived who did both with greatness. That is lucky for sure.

His father, who was the police court judge at the time of Tom's arrival on earth, celebrated the event by releasing a jailful of prisoners—mostly bums and drunks. It caused quite a problem for the policemen who had to round them all up and rearrest them, but Tom's birth had been uniquely celebrated.

Tom's mother taught piano lessons even before she married his father. His father loved Mexico and would go there to explore old mines, adventuring in every town. Tom's father had some cousins who had a ranch adjoining the Three Rivers Ranch (owned first by Albert Falls and later by his friend Charles Leavell). He had gotten his license to practice law in Missouri, but hunted and cowboyed with his cousins before arriving in El Paso with one dollar in his pocket and not knowing a single person in town.

Tom Lea, Sr. went on from judge to be elected mayor of El Paso in 1915. The Mexican Revolution was in full swing at the time. His father had to sort of keep the peace on this side and their part of the border. Once, when Pancho Villa crossed over to El Paso, he and the mayor had some words. Villa didn't like the mayor at all because he had ordered Villa's wife, Luz Corral Villa, and Villa's brother, Hipolito, to jail for smuggling arms. So the famed revolutionary put out a reward notice of $2,500 in gold for Tom's father, dead or alive, and sent a threat to kidnap young Tom and his brother Joe. The latter survived to contribute enormously to Charles Leavell's vast construction companies and the former to the art in his soul. Already, way back then, the great web of destiny was being woven. At any rate, young Tom, Jr. would look out the front window of their home and see a policeman out there for protection. They attended school with a police escort during the revolution. Heady adventure for such young men!

Tom became editor of his high school paper and studied art with a gifted teacher, Ms. Gertrude Evans. In 1924 he headed back east, with his family's blessings, to study at the Chicago Art Institute. His mentor there was John Norton, whom he has appreciated and thanked all his life.

Even with art study in Europe and all his other travels, El Paso would always be the first home and Mexico the second home of his soul. Al Lowman of the Institute of Texan Cultures quoted Lea in October 1969 as saying, "Those of us who love both banks [of the river] wish there were more and better bridges—we wish more people traveled them with their hearts."

Tom met J. Frank Dobie and out of that came his acclaimed illustrations for Dobie's books, *The Longhorns, Apache Gold and Yaqui Silver* and others. During this period he made the single greatest move of his life. At a party in El Paso, he met and later married a young woman from Illinois. Although he proposed to her the very next day, she made him wait a few months before saying yes. Sarah Dighton was and is a classic beauty. She fit right into El Paso society, becoming a civic leader and taking over the financial side of Lea's career, but also becoming his main confidante and critic.

At this time Lea was also winning mural commissions across the land, including one for the Post Office in Washington, D.C. Very near the same time that Charles Leavell was launching his career by constructing buildings and runways for men and machines, Tom got a wire from the editorial staff at *Life* magazine.

He says simply but succinctly, "I bought a brand new sketch pad, and in the fall of 1941, I went to sea aboard a U.S. Navy destroyer in the Nazi-submarine-haunted North Atlantic as an accredited war artist correspondent of *Life* magazine. The next four years were a huge break from work in my cherished corner of homeland."

Like Charles Leavell, Tom Lea contributed to the war effort according to his special areas of expertise. Tom adds, "I want to make it clear that I did not report hearsay. I did not imagine or fake, I did not cuddle up with personal emotions, moral notion, or political opinion about war with a capitol W. I reported in pictures what I saw with my own two eyes wide open."

What he saw and felt in the frigid North Atlantic was the terrible stalking of killer submarines sending Allied materi al and men to watery graves. He drew it and noted it in all its frightful destruction of steel and nerves. The men of the destroyer respected and loved him as one of their own and he reciprocated those feelings.

Tom Lea toured the North Atlantic, the South Pacific, Europe and North Africa, as well as China theaters of operations, to tell his unsurpassed pictorial truth, along with other combat artists such as Bill Mauldin (born a few miles north of Lea, in High Rolls near Alamogordo, New Mexico), William Draper, Edward Brosney, Hugh Cabot, and a few others. The publishing of his drawings and paintings of the invasion of the Island of Peleliu preserved forever the true sadness of war, revealing it in all its horrendous bloody chaos to the American home front. The wake-up shock waves across the country were great.

He waded ashore with the First Marine Division at dawn on September 15, 1944, right into as bloody a hell of ravaged earth and flesh as could be imagined. The Japanese waited until they were in the cross-fire. This border artist of such classic tenderness and love was in an earthly Hades of inescapable, deafening artillery explosions, the sharp rattle of machine guns and the singular crack of rifle fire breaking the sound barrier overhead or thudding shatteringly into flesh and bone. Smoke and dust and blood filled the air with an indescribable smell that is unforgettable to all who have entered this particular inferno. Some twelve hundred marines died and thousands more fell wounded here, all around and next to Tom Lea.

One blood-soaked casualty came staggering up and fell right next to Tom and another wounded Marine, Bill Tapocott, who would live to verify the event exactly as Lea depicted it. *The Price* is the most controversial work Tom Lea ever did. He wrote,

> Mangled shreds of what was once an arm hung straight down as he bent over in his stumbling shock-crazy walk. Half his face was bashed to pulp. The other half bore a horrifying expression of abject patience. Grotesquely his blood-soaked uniform was coated with coral grit. . . . He never saw a Jap, never fired a shot.

The Two-Thousand-Yard Stare was the painting for which Lea became most famous. It shows a Marine staring blankly, straight out of the picture, looking away from Bloody Nose Ridge, the last, strongest and most terrible fortress on the island. The shell-shocked face goes straight to the viewer's soul and tells the world of war's inner ghastliness like no other painting since Picasso's *Guernica*. More than two-thirds of this Marine's company had been killed or wounded; how much more can he endure? Certainly these works will endure as long as anyone can see or care about their lesson to all humankind.

It was during Charles Leavell's critical home-front war efforts, which met with such substantial skill and success, and Tom Lea's return to El Paso from far and deadly voyages, that they met and became friends. More than friends—they were comrades in the seeking of beauty. And their beautiful, dedicated and talented wives joined them. The foursome was as rare as a pound of emeralds and their friendship even more precious. It has lasted more than fifty years and will forever.

It was a reciprocal friendship in all ways. With words and brush, Tom created permanently the land that Charles loves and the people he both loves and respects. Charles became Tom's biggest fan and no doubt owns more original Leas than anyone else on earth. Both Leavell and Lea were givers to their friends, their families, their country and much of the world.

In 1951, Charles Leavell introduced Tom Lea to his uncle, John Leavell, a real rounder who had won the Distinguished Service Cross for Bravery in World War I. Tom said, "He was one of the old boys who helped shaped my life. He had even been a whaler and later driller and found lots of oil. So he told Charles and me we didn't know anything and needed to learn. He took us on a plane to Rock Springs, Wyoming, and there introduced us to Clem Skinner, outfitter and packer, who had been his sergeant in World War I. Clem stopped off at most of the bars between Rock Springs and Pinedale. He knew all the barmaids. We spent a couple of weeks at or near Horseshoe Lake in the Windriver Mountains riding, fishing all the great streams and lakes up there. This trip brought Charles and me very close together."

Soon thereafter the two decided to take their wives up there, and they continued this annual vacation adventure for many years. Charles relates, "Tom was a fine rider, he'd learned when just a kid. He'd ride into the high country and never get off his favorite Clem Skinner horse named Lagrima—it means 'tear' in Spanish. He'd just fish right from [the horse's] back in the roughest places—and did a damn good job too. He did a painting of a great waterfall—named Skinner Falls—that we camped near, and it's in my office in El Paso. I look at it every time I walk in that room and remember the great times the four of us shared together in that mostly uninhabited part of the wilderness in those glorious days back then."

One early evening, all four friends met on their favorite table rock to celebrate a priceless day of fishing and companionship. Clem made them martinis chilled by snow from a nearby snowbank. The moon was coming up and the mountain breeze was just enough to waft all the delicate scents of the pines and spruce across their area. A coyote howled and was answered by another starting an early hunt.

They visited awhile and then suddenly Tom and Charles made a vow. They both swore that no matter what happened in worldly fortune, El Paso would always be their home. Shirley and Sarah eagerly joined this toast to the border desert while at the same time enjoying the Wyoming mountains. And they are all four still in El Paso. A timeless, precious moment, that.

While Tom Lea was writing and illustrating the massive, two-volume King Ranch books commissioned by the ranch owners, the Klebergs of Kingsville in South Texas, Charles often accompanied him to the Klebergs' Argentine ranch, where they adventured and fished the Andes of Argentina together.

Mary Lewis Kleberg's *Salute to Tom Lea*, from May 4, 1990, echoes the feelings of the Leavells for their friends Sarah and Tom Lea:

> My heart overflows with memories of his many wonderful days in Kingsville with family and friends, our trips to ranch properties in foreign countries. He rode in hunting cars in the pastures of South Texas, rode in Land Rovers in Australia and saw Blue Ridge dingoes rounding up cattle in the "down under," rode horseback on the Pampas of Argentina to learn the gaucho's way—all of this so that Tom could absorb the aura and substance of what was King ranch and its people.

Yes, Tom Lea is beyond compare in his combination of quality and range in both writing and painting. His art is endless and vast in its powerful and delicate reaches. He paints or illustrates in any medium or style, one as fine as the other. He does portraits, stark realism, impressionism, cubism, graphics, oils, watercolors, ink and casein. His novels, such as *The Brave Bulls*, *The Wonderful Country* and *The Hands of Cantu*, are poetically powerful, with a deep feeling for and skilled knowledge of the wild desert land and all its living creatures he loves and portrays so profoundly. (*The Brave Bulls* and *The Wonderful Country* were made into movies; the latter, starring Robert Mitchum, is considered a western classic.)

Lea recently said, with absolutely no rancor, "I'm not well-known at all outside Texas, and maybe parts of New Mexico," but this is very difficult to believe. Charles Leavell, in a speech in honor of Tom Lea at the El Paso Historical Society, said this: "Tom, as an artist, describes our land as 'a thirsty, bare and mostly empty country. It is tan, not green. It has no abounding grace of fertility and little softness to evoke ease in man's spirit. Its richness is space, wide and deep and infinitely colored, visible to the jagged mountain rim of the world—huge and challenging in space to evoke huge and challenging freedom.' "

CHAPTER 21

Seeking the Lost Lakes:
The Demolishing of Burdens

The Leavells set up an airstrip with a hangar near Creede for their twin-engine Lockheed. Often Charles would have the pilot fly over parts of the 4UR realm just for the pleasure of viewing its majesty. Early on he spotted two small lakes up in the high country that were like diamonds set by the gods when the sun sparkled on them just right. They intrigued Charles. Somehow he felt they should be part of the 4UR. He knew there would be many guests who would love to pack in, camp out and fish such wondrous high-country water. This would wrap up his dream for the 4UR.

Now, Charles is a generous, loving man, but—as many can attest—he can be cold-blooded about acquiring or accomplishing something he believes in. So he decided to add these gems to the 4UR jewelry box no matter what it took. First he found that the family that owned the lakes had split up and moved to Seattle and other parts of Washington. The main point, though, was that the property was now available.

The "Lost Lakes" were at an attitude of 10,400 feet and only pack horses and mules or hardy hikers could make it to them. Charles hired a helicopter and flew right up to them for a real look. They were even more than he expected. The views and the wilderness set his soul singing, but when he pulled a huge native trout out with a fly for the first time he became obsessed with possessing this treasure of the Rocky Mountain wild.

Charles Leavell.

After exploring around in a four-wheel-drive vehicle, he found a way up there by land. In the meantime he had made a bid that was accepted and contracts were drawn up for sale of the private land. He then sold all the valley farmland attached to the Lost Lakes property for enough to pay for the lakes. He now had 1,110 acre-feet of water—one lake contained 75 surface acre-feet and the other 35. The lakes were important indeed, for they were part of the headwaters feeding the Rio Grande River. Besides supplying two miles of great fishing on the eastern edge of the 4UR and beautifying Creede, their added liquid gave the land a bountiful eighteen-hundred-mile highway of water all the way to the Gulf of Mexico. On the way down there, it watered rich crops of produce, orchards and farms, livestock and wildlife and gave life-saving nourishment to Charles's beloved border desert, before it went on through the Texas Rio Grande Valley and on into the Gulf.

True, it was in a truly breathtaking setting. True, it was full of native Colorado cutthroat trout. But there were two or three parts of the National Forest that touched the lakes. This made them accessible, in these places, to the public. That meant people would read the maps, walk up there, fish, skinny dip, picnic and leave their trash all over the place. Much disrespect had been already been shown to the land. Charles tried to justify the public's right to go there, but no matter how hard he and his helpers tried, they could not keep it litter-free and clean. It just proved impossible. They were offended at the cluttering-up and trashing of such a lovely spot. These lakes were now part of Leavell's long, long trail to the 4UR paradise and this defilement was sacrilegious to him. That is how he felt and, as always, he acted upon his personal conviction.

Charles made friends with the forest supervisor in Monte Vista, Colorado, and told him, "The Forest Service owns some land at Lost Lakes and I want to buy it." The supervisor explained that it was against the Constitution to sell it, so they compromised. Charles bought several pieces of private land adjoining or in the middle of other National Forest sites in Colorado and Wyoming. Then he traded those to the Forest Service for the land at Lost Lakes. These trades allowed the Forest Service to consolidate holdings that long inconvenienced them and Charles now had Lost Lakes protected for his guests from around the globe. It has been used with cleanliness, dignity and grand adventure ever since.

They improved the hidden road just enough to get in with camping equipment so special guests could experience the high wild country and fish for the big native trout. Soon they built a small rest house up there and put in some bunk beds and a stove. No one is allowed to go without a professional guide. Charles says, and means it, "I don't want anybody hurt."

No motorboats are allowed. They have two quiet electric boats with a guide on each one. The creatures of the forest have it all to themselves most of the time and there are no loud, grinding motorboats to frighten and insult their delicate sensibilities. It had taken skill, and much jaw-clenching dedication from Charles, but the 4UR was now complete, and in his mind a Shangri-La, as a letter to the two things he loves the most, Shirley and the 4UR, attests. The letter was published in a very fine fishing book, *The Fly Fishingest Gentlemen*, by Keith C. Russell:

Charles Leavell fly fishing on Walton Pond with the Rio Grande Palisades in the background.

ONE INFINITE MOMENT

Shirley,

Do you remember our search for the fisherman's Shangri-La? Shirley, I don't think that Shangri-La is so much a place as it is a state of mind.

For I vividly recall sitting in our boat late one evening when the sun had gone behind the divide, and the cliffs were mirrored in the dead calm Lost Lake. We were casting under the half-light and found ourselves afloat in the clearest water I had ever seen. There was no definable horizon line as air and water became one. I remember that a great sense of well-being flowed over us as we felt suspended in time and space.

Suddenly, a tremendous school of bait fish swept under us as though swimming in air, and a school of rainbows was slashing through them, leaving silvery scales to settle from the fury of their attack. Then, mysteriously, underneath it all we saw an immense form, the shape of a trout so huge that he had to have been close to a century old.

But before my fly could settle to him, a second trout hit as I tried to take it away. While I was still hung into this smaller fish, the silver school and the great fish moved off into the mist. It was then you told me, for one infinite moment, together we had seen and felt the mystical Shangri-La."

CHAPTER 22

The Daily Workings: Fulfilled Yearnings

Finding the different people connected to the 4UR, who do the planning and create the attitudes that go into making and preserving the uniqueness that was always there, has not been easy either. The whole point, however, is to make it seem so for the guests, while at the same time making the employees and owners feel comfortable and at home—all part of the 4UR family. In return, they get peace and pleasure untold, having given it to others. Amazingly, this is accomplished here as in few other places in the world.

One of the principal reasons Charles Leavell has been such a success in his worldwide building endeavors is his ability to choose, or inherit and keep, the talented people who have total loyalty to performance of their duties. The legacy of Ed Wintz, for example, was beyond price in value to Charles and the 4UR. Bill Geis, the master trout fisherman, was no less a treasure of skill, diplomacy and loyalty. They are held by all in the greatest respect. Also, without exception, every one of these people naturally speaks of the 4UR in almost reverential terms like: "This is a special place, all right." "Yes, the 4UR is very special." Special, special, special. They all say it, and they all know it, each for their own special reasons.

The young ranch managers, Rock and Kristen Swenson, are solidly in the 4UR-lover's camp, and their admiration and esteem for Charles H. Leavell is limitless. A man as strong-willed and accomplished as Charles makes enemies—and a few may be justified.

Rock and Kristen Swenson, the current managers of the 4UR, at the dinner bell.

However, most of them come from jealousy. It's natural. These petty people must suffer greatly at the solid backing he gets from his associates.

Anyway, when he hired the Swensons a decade ago as ranch managers, he caught the two biggest trout in Goose Creek. They were experienced from having been assistant managers on a large guest ranch in Wyoming, the A–A. They work together now like an all-pro quarterback and his favorite receiver. From the beginning, Leavell knew they would be involved in every aspect of the ranch: keeping books; making reservations; overseeing the chefs and the dining room, waiters and waitresses; bartending; tending the livestock; "wrangling" or guiding the guests on horseback rides of viewing and exploration; and, dauntingly, a lot more. Both the Swensons are good with horses, but one of them would have to take the horses as an individual responsibility and the other the fishing waters.

To help make the choice, Charles took Kristen fishing with him on Goose Creek. In the first hour she hooked his hat several times. The decision was made for him: Kristen would hire the ranch foreman and be in charge of the horses. Rock would handle the creek waters—and, of course, they would both take care of countless other items to make the ranch run smoothly. It is amazing to watch them accomplish this. They both seem to know instinctively what is right. Their own confident skill and knowledge makes these jobs look easy; it's tremendously deceiving. It all seems so effortless! It takes a form of dedicated magic to create this relaxing flow of events. Their love for the place is obvious in their care, but in their talk you hear it for certain. They are both people of such dedication to the 4UR, and their souls are so deeply embedded here, that presently one cannot think of the ranch without their provident presence.

Rock says his main desire for the 4UR is to see a continuation of the tradition of improving the land, the waters, and the wilderness for the up-coming generations of guests, to make them even better than they have been in the ranch's long prior history. Kristen smiles introspectively as she tells about a little girl on the horseback trail who for the first time sees a young elk raise up out of tall mountain grass. She says, "The wonder in her eyes at confronting pure nature is so beautiful you can't describe it."

It is a blessed sight to watch wranglers riding in the lead of a string of good horses, bearing wide-eyed children, knowing that very soon they all will be safely, expertly right in the middle of nature's wildness and beauty. On their return, they may be tired almost to exhaustion, but their eyes still see the rarefied greens of the forest, the rich browns and reds of the great upthrust rocks and all the creatures that fly about the bushes and trees. Now they know where the animals rest, graze, hunt, love and take care of their families, just as the firm but gentle cowboys have cared for them. The benefit to their young hearts, minds and spirits is boundless.

In the late winter and early spring, the 4UR sends out applications to selected colleges around the country for the season's help. These are screened carefully, as the 4UR people want those who understand and appreciate the family atmosphere so long nurtured here. They get hundreds of replies. And why not? Anyone wanting to have a well-rounded education could hardly be employed at a better place than this guest ranch. Many, especially those in the hot and humid South, are eager to have a paid working season in the cool altitudes of the Colorado Rockies. They work as waiters, kitchen help and outdoor maintenance assistants, as well as desk labor, babysitters and temporary tutors.

Years ago Charles invented a spring party to break in the help. They have about forty guests putting on skits. They have competitions in fly casting, fishing, and horseback riding. In the evening there are dances of all kinds, including square dancing. As Charles says, "The necessities of life."

The young people love it and get the feel of the place, inside and out, right off. A great number of those accepted come from colleges in Virginia and the South because, Kristen says, "They are so polite. They say 'Yes, ma'am,' 'No, ma'am,' 'Yes sir,' and 'No sir' with old-fashioned politeness. It's something they have grown up with, and they're comfortable because it's natural."

The 4UR boasts a guest list of forty-year returnees and a few who have been coming there for fifty years. Charles speaks affectionately, "We've got one old boy up here by the name of Wilcox. He's from St. Louis. I sure am fond of him. Just this last November he had his seventieth wedding anniversary and he told me, 'I took my little bride up to the 4UR right after we married and we've been coming back ever since.' He's over ninety and his wife is eighty nine. He's a tough old bird, and I mean to tell you he loves that Goose Creek water."

Charles continues, "Some of the guests are so at home here they often feel they are not only part of this elongated, wide-spread family, but they have owners rights as well. They tell me how to keep the whole thing going and they don't want me to change anything while we're improving it, especially the rates."

A preponderance of the 4UR guests are, naturally, from nearby Oklahoma, New Mexico and Texas, but they also hail from New York, Kansas, Michigan, San Francisco and all over. There have always been a smattering of foreign guests. Charles describes one sample, "Two years ago we had a man from Wimbelton, England. He called himself Sir Tits, but he said, 'That's not my real name, just my main interest.' He was a dandy. Another man returns every year from South Africa to

Shirley on her pacing mare near the 4UR.

spend a few days fishing, eating, drinking and absorbing the mineral water, the mountain scent and spirits into his blood. He calls his trip 'a treat for all time.' There is something here, something not quite definable, a magic you see and feel in the eager, but relaxed guests no matter what their ages."

This magic shows as well in the management and in the energy and contentment exuding from the college help, no matter what their assignments. It's present in the horseback riders heading out and returning; the fishers doing the same; in the tennis players and the trap-shooters; and in those who love both the swimming pool and the hot springs. Everything turns and moves as smooth as the essence of a single silken soul. That is not only amazing, but also, as all have said, very, very special.

The 4UR is an escape from the rigors of the outside world—a total change of pace, a freeing of the mind. Activities are planned for those who wish to participate: hiking, jeep tours of the mines, ghost towns, Creede and other historical and wilderness areas. Box lunches

Launching a float trip at the 4UR with Lookout Mountain in the background.

are provided for the tours and special picnics and cookouts are supplied, including one noon fish fry and a chuckwagon breakfast weekly. If the pace is too fast, some people prefer to just lie back and read a book, and pause now and then to look at the soothing mountains and breathe the fresh cool air, or hustle up a bridge game at the lodge where there is a game room and a colorful bar.

There is a well-supplied, self-serve ranch store with fishing gear, a large selection of flies, varied apparel, caps, tennis balls and sundries—all run on the honor system. No clerk is there to help. The buyer leaves a note listing whatever is purchased and pays for it at the end of the stay.

There are public telephones near the store and main lodge but none in the rooms. Any emergency or after-hours calls are relayed to guests by the management. The mail is picked up in Creede daily, except on Sunday, and delivered to the guests. Family movies are shown each Wednesday, complete with hot popcorn. Laundry service, either by the staff or do-it-yourself, is available at the pool area. Escape from the responsibilities of the world outside, convenience and choice make up the menu here.

Trout from Goose Creek.

4UR Pierce Creek breakfast ride.

Goose Creek and a portion of the 4UR headquarters, with the famed Wagon Wheel Gap palisades more than three miles distant in the background. (Courtesy, Pat Evans)

To top off a great day of fishing or adventuring, there is a sauna room and hot mineral pool available in the bathhouse. The hot pool, at 104 degrees, provides soaking pleasure for all who wish it. On Fridays, a licensed massage therapist, Grace Russell, is available by appointment.

Counselors provide special programs for children five years or older. The youngsters also have available lessons in riding, horse care and grooming, with real cowboys for teachers.

A massive bell on a rock stand just outside the lodge signals the beginning of each meal. The food is absolutely gourmet. The daily choices of meals are posted on a small blackboard at the dining room entrance. There are no menus. (It is astounding how this approach cuts down on clutter and all-around bother!) However, if a guest does have a craving for something different, it can be arranged with the chef. The dining room staff take great care in pleasing.

Everyone still occasionally talks about the visit of the renowned gourmet chef and television personality, Julia Child, and all the special

Knee-deep grass in a 6.5-mile-long 4UR meadow. Forty horses (guest and personal) graze here year-round. Uncounted deer and some two hundred to four hundred elk live abundantly here through much of the winter. (Courtesy, Pat Evans)

preparations the chef made for pleasing her. With the news of her reservations for a large retinue, including her business manager and a niece and nephew, the head chef was at first in ecstasy and then the very opposite. The nearer the date of the great one's arrival, the more anxious he became. His chef's imagination went on double overtime. He had heard she was interested in Indian food, so he researched menus, bought special ingredients and cooked and cooked to achieve perfection. He also planned all sorts of gourmet dishes that might please the selective palate of one of the world's most famous chefs.

Mr. Leavell smiles with both his face and eyes as he says, "She finally arrived, to the relief of everyone. And you know what she wanted to eat? Fried chicken and hamburgers. She never had a bite of Indian food. Never mind any of that, however, because she did fish and was pretty darn good at it." So in the end, she passed the main man's test.

Just to prove how unpredictable people are—and also to prove that the chef's concerns were not unfounded—a recent New York press

Jimmy Barnard, a regular guest for fourteen years, seeking "the big one" right next to headquarters. (Thirty minutes after this picture was taken, he landed two.) (Courtesy, Pat Evans)

Jack and Daphne Cheatham select flies for a day of world-class fishing. (Courtesy, Pat Evans)

231

release on Julia read, "If they ever executed Julia Child she would break the bank. For her last meal, she would order 'foie gras oysters and a little caviar to begin with.' TV's favorite chef told *the New York Times* magazine in the April 2nd issue, 1995, she would follow that with pan-roasted duck and her beverage of choice would be a 1962 Romanee-Conti, which sells for $700 a bottle and with dessert she would have Chateau d'Yquem, 1975 or 1976, that's $450 a bottle." One suspects that Julia will escape execution and enjoy these little repasts as she damn well chooses—although at a certain second of a certain day on the 4UR, there was probably a chef who would have gladly signed the final warrant, even though he has had many laughs about his experience ever since.

Walt Disney filmed *Cougar Country* on the 4UR. He loved the place so much that he stayed on several days after the final wrap. The cougars still roam there, but because they have such a vast area of forest land in which to hunt and survive, the naturally wary cats are very seldom seen. Those with experience often observe their tracks, though. They are part of the unseen, but deeply felt, presence of the place.

If anyone at all should think that the praise of the 4UR in this journal is overabundant, please think again, and listen to words from a few more of its current guests.

Ralph Cousins's father Bob first brought him to the 4UR forty-five years ago, when Ralph was only three years old. He has been there every summer since. Cousins says earnestly, "I recently formed a new company and named it Goose Creek, Inc. If I have a worried, sleepless night coming on, instead of sheep I count trout I've caught in beautiful Goose Creek and soon I'm resting peacefully." He and his wife, Harriet, bring their sons Christian (age sixteen), Robert (age eleven) and David (age nine) with them every year, and the boys love the 4UR just as much as the generations before them. Continuity assured.

Bill Wise, his charming wife, Marie, and their daughter, Genna, have been guests at the 4UR for nine years. Now, Bill Wise is a man of great business acumen and has the wealth to prove it. His family, as so many other 4UR guests can and often do, travels the world, including African safaris, trips to Paris and Rome, sojourns on the French Riviera, and anywhere else they please. However, they don't return to Paris or

Ralph Cousins started coming to the 4UR 45 years ago with his parents, Bob and Dorothy. Now he brings his wife, Harriet, and their three sons to carry on the tradition. (Courtesy, Pat Evans)

Christian, David, and Robert Cousins warm up in front of the lobby fireplace. The temperature on this June morning is about 40°F. (Courtesy, Pat Evans)

233

Bill Wise (left) and his good friend, Jack Cheatham, prepare for a prime day on Goose Creek. They and their families have been guests at the 4UR every year for the last nine years. (Courtesy, Pat Evans)

any other spots in the world—wonderful as they may be—every year; but they can't wait to get *home* to the 4UR with absolute regularity.

Accounts such as these are constantly being given, without any prompting, by all living generations of 4UR workers and guests—that is, the family.

CHAPTER 23

Mines and Moguls: Saviors Come Calling

Firstly, the 4UR realm was a mighty hunting domain for the Ute Indians and many other tribes. Secondly, it was transgressed by plains Indians. Thirdly, Spanish and Anglo explorers came looking for new trails to the Pacific Ocean and a source of possible riches. All the latter were found, but it was the gold, silver, and other more crucial minerals that caused a great romantic boom, creating overnight moguls, gambling and drinking madness, gunfights and ambush slayings and the raping of the land, and created the power of wealth and fame that went mostly to other parts of Colorado and to magnates inhabiting eastern cities—many of whom never saw or suffered the source of their wealth. The 4UR Ranch location almost fell prey to this unthinking, hypnotic madness that has forever infected the blood of humankind.

Around 1891, during the time of the great boom camps, the first claims were staked at Wagon Wheel Gap and the surrounding area. Because some ore had the same coloration as product of the famed and very rich Amethyst Vein at Creede, it was mistakenly believed to be an extension thereof. The owners of the claims were, naturally, extremely excited. They fully believed they would quickly pocket the same enormous wealth as had come from the Amethyst, but assays soon showed it to be fluorite, mixed with barite. There was no market near enough to make it profitable to mine, but its time would arrive. Before that, several different people owned the mines, betting on the future. Charles H. Leavell would one day own the silent mill and earth it sits on.

However, in 1911, a mining engineer from Creede, S. B. Collins, recognizing its value, purchased the property from claim holders. He organized the American Fluorspar Mining Company. From 1911 to 1921, the mine was worked spasmodically and produced forty to fifty thousand tons. Most production—fortunately for the 4UR guest ranch—came from upper workings, and an aerial tram operated by gravity carried the ore to the mill at Goose Creek, where it was concentrated in a jig of commercial grade. At first, the concentrated ore was hauled by mule-drawn wagons to the Gap, where it was loaded onto railroad cars. This was finally proven inefficient and a tram track was laid alongside Goose Creek so that the mules could pull a number of loaded cars to the railroad. It makes one wonder if there would ever have been a West—and most of the rest of the early explored world as well—without the help of mules.

The C F & I mill in Pueblo, Colorado, used the metallurgist-grade concentrate for flux in its open-hearth furnaces to make steel. The huge Pueblo company shipped the resultant iron and steel across the continent to help build it. The 96-percent grades of concentrate were shipped to St. Louis and to Illinois to make hydrofluoric acid.

In 1913, engineers took an interest in the mine's mineralogy. During a close examination, they discovered a new mineral known nowhere else and named it creedite. The 4UR is unique even in its minerals!

Colorado Fuel and Iron purchased the property in 1925 and took over the mining, milling and management. The old mine was abandoned and three levels were developed: the New Tunnel, the Intermediate Level and the Wilson Level. C F & I operated the mine until 1950. In 1980, the Leavell family purchased the mill and all surface rights.

To step inside the three- or four-story-high mill is to be awed by the massive array of engines, rock crushers, huge ropes and belt pulleys, and wonder how this tonnage of working steel was ever moved here at all. But when a fertile imagination starts the pulleys turning and the great wheels and engines grinding, the conjured-up noise and the

Guest horses grazing in front of the old fluospar mine and mill.

shouting of the grimy and sweating mill hands is deafening. Then there's a sudden realization that most of the glamour from mining came only from the initial thrill of discovery and to those in banking and investments who finally got the checks for the finished metals, delivered by shattered bones, blood and burial. All the rest of it was wearing, crippling bewilderment. Nonetheless, the towering, rusting mill abundantly adds to the overall mystique of the 4UR Guest Ranch, preserved by the Leavells as a reminder of what went before in this beautiful, peaceful valley.

CHAPTER 24

Genes of History: The Family of Perpetuation

In a time when scientists promise many miracles from the reordering and experimentation with genes, one wonders if the little rascals shouldn't be left to work everything out on their own. It seems their doing so has added to the enormous diversity and often terrible, often wondrous, activity of the human animal. However, the genes of Charles H. Leavell and his wife, Shirley, certainly point to a genetic destiny.

Charles's maternal grandfather, William Wallace Walton, fought through the entire Civil War with Company G of the Second Texas Infantry. He was in many terrible battles, including forty-four starving days at Vicksburg. He was wounded seven times but, with amazing fortitude, finished the war right up to the surrender. Though he never talked of his own heroics, he did recount in writing this short segment:

> I was near Albert Sidney Johnson when he was killed at Shilo. Was with Col. Bill Rodgers of Houston when he got killed at Corinth. We charged at Ft. Roberts and got in behind trapped. Col. Rodgers thought we were doomed and he tied a handkerchief on his sword and tried to surrender us, but the Federals ignored the flag and fired on. Col. Rodgers turned to us and said, 'Boys, I'll never give up now. Dying is probable.' Then he turned his horse to the right and made him jump a rifle pit, went riding in, emptying two six-shooters before they killed

him. Then the Federals propped him up against a stump and took his picture, as they counted him such a heroic man. Those of us living were taken prisoners but many of us soon escaped to serve on.

I went in at the first of the war and stayed until the last, but want no more war for me or mine.

After the war he moved to Davilla, Texas, and started a successful mercantile business in 1870.

In World War I an uncle, Captain John H. Leavell of Company F, 316th Combat Engineers, was awarded the Distinguished Service Cross by General Pershing for leading a patrol of four men into the city of Andenarde, Belgium, on November 1, 1918. Though badly outnumbered by a German patrol in the combat that followed, they killed many of the enemy and captured an important spy. The Leavells were reluctant but good warriors.

Another uncle, Jimmy Holland, heard about the Sawdust Swindle: selling $10,000 worth of fake money for $500. This swindle had moved all the way from hurting people in Texas to hurting people in New York. Uncle Jimmy Holland decided to stop them in the big city and make them pay for damaging his friends and family. He went to New York. As the Davis brothers tried their scam in a New York apartment, Holland put the real money in a black bag before they could make their switch. Ted and Tom Davis screamed for him to stop because, as they weakly explained, detectives were all over the place. Ted yelled, "Tom, kill the son of a bitch." Whereupon Tom reached under his coat for a gun, but Jimmy was faster at pulling his own gun and shot Tom dead with one bullet. Holland spent six months in the Tombs while readying for trial. All the New York papers became full of the 'Texan who had come to rectify past sins for his fellow southerners.' The papers loved it and old clippings show such headlines as: "THE TEXAN—HOLLAND HAS FRIENDS. Holland's [Uncle Jimmy] shot the famous `Green Goods' trial now begun."

Well, he certainly did have friends. Among those coming up from Texas to support him were his brother-in-law, Civil War hero, John H. Leavell, Sr.; Colonel Frost of San Antonio; Henry Lamar of Georgia (a cousin of President Lamar of the Republic of Texas); Chief Justice

Marx of Texas; General J. G. Tracy of Houston; and a horde of others, who formed a protective cordon around Jimmy Holland in the courtroom. The trial moved on, with both prosecution and defense fighting hard, and the papers printed almost every word.

The prosecution had Uncle Jimmy demonstrate his ability with a gun in the courtroom. He stood, arms loosely at his sides, with a Colonel Fellows playing the dead Davis. "Well . . ." began the prosecutor. Holland whipped the pistol out so neatly and quickly that Fellows jumped a foot in the air. It was written that a cricket would have been proud of such agility.

The trial continued, but on February 29, 1886, a New York newspaper headlined. "THE TEXAN ACQUITTED—only fifteen minutes required by the jury." Uncle Jimmy was complimented all around, but one high-collared man said, "I don't want to be heartless, but if you had only killed both brothers this city could well afford to make you a present. You have earned the respect of all honest men."

Uncle Jimmy returned to Abilene, where his land and cattle business flourished. In 1887, he became a resident of Amarillo, Texas. The *Amarillo Star* of May 6, 1904, said this upon his death:

> Mr. Holland and H.B. Saunders laid out the present city of Amarillo. He was one of the founders and directors of the Amarillo National Bank. He owned considerable real estate in Amarillo and a ranch in the country. By his business ability, he acquired a considerable fortune. He was most charitable and gave largely to the poor and needy. He was interred in the Odd Fellows Cemetery alongside his sister Mrs. John H. Leavell.

His strength of character, combined with what we know about Charles's father, gives a powerful argument to those who argue that genetics— or breeding—will tell.. The Leavells' blood and kin have been strong, and doers of good, right until today.

One could make a list of Leavell go-getters as long as the Santa Fe Railroad, but at least one more must be mentioned because he became a "border" man and a damned important one. Charles's great-uncle, Tom Holland, who married Elvira Boozer, was the engineer in charge of surveying the eighteen hundred miles or so of border between Mexico and the United States in the years 1856 to 1881. His was a challenging job, to say the very least. In this harsh land of singing sun, flash and drowning floods, various poisonous serpents and insects and abundant roving bands of outlaws he created an underrated accomplishment that was done in high order nevertheless. From Brownsville, Texas, on the Gulf of Mexico to San Diego on the Pacific is a border of mighty significance in this country's history, and it will only increase in commerce and contradiction as the world moves on.

Charles wife, Shirley (of the Terrells) has genes as powerful and varied as those of the Leavells. Her family came by way of Essex, England, to Virginia and Georgia. Shirley's great-great-grandfather, William Terrell (1732-1793), received bounty land in old Wilkes Country, Georgia, for Revolutionary War services. Shirley's line of the family wound up in Texas. The Terrell men were mostly doctors of medicine. Enough of them settled in South Texas that the town of Terrell was named after them.

Edwin Terrell, a younger son of Dr. Alexander Terrell and wife Sofia, died in 1883, according to his tombstone at Hempstead, Texas. He was a member of Company B, Eighth Cavalry, under Captain John Al Wharton. He served with distinction, and in 1869 he married Katherine, daughter of the famous general, Richardson Scurry. This same Edwin Terrell later became a railroad official and contractor who built the Texas Central and the Texas Midland Railroads. Again, there are so many parallels, not only to the Leavell family, but also to the first owner who developed the 4UR, General William Palmer.

We are aware of Shirley Leavell's help and total dedication to Charles's career and their resultant close-held family of blood and friends. But what hardly anyone knows is her silent and selfless service to the development of the El Paso Y.W.C.A.—now the largest in the United States—and the Junior League. She was one of the first two women on the United Way board. She was also director of Family Services, the Boy's Club, the Museum Guild, Goodwill Industries, and dozens of other charitable organizations and projects. She has received so many awards over the years, from all faiths, for her services to the underprivileged that it would be, by her true and simple good nature,

Charles and Shirley Leavell.

embarrassing to list them. Perhaps she wouldn't mind so much people knowing that her work with the Y.W.C.A. exploded into the national scene with a cover story in *Parade* about America's first residential intervention center for predelinquent girls; that was quickly followed by the national bicentennial award for one of the largest child-care programs in the huge state of Texas. It really does seem that her placid doing of these honored deeds would be joined by Charles. He was, just as she was always there during his often-daring achievements.

Whether the gene theory holds true in all cases, one must admit that the bloodlines of the Terrells and the Leavells have held true right on up to this immediate generation. Both have always been families of great vitality and often of unintended, overwhelming presence.

The Leavell family, Christmas 1994. From left to right, Kelly Pinkerton, John Pinkerton, Charles Leavell, Tip Pinkerton, Brian Leavell, Lindsey Leavell (seated), Fred Winston (seated), Mary Lee Pinkerton, Matthew Winston, Shelly Winston (kneeling), Pete Leavell, Shirley Leavell.

In 1960, the year in which Mary Lee Leavell was named Sun Carnival queen, a young reporter, Brian Woolley relates:

I was eighteen years old—the greenest cub reporter that the *El Paso Times* or any other newspaper had ever seen. I mean I was RAW—6 foot 3, 135 pounds, and brand new into the city from the Davis Mountains. I was a full-time student at Texas Western College (now UTEP) hanging out in Juarez a lot and getting very little sleep.

Normally, such an interview would never have been assigned to Woolley, but that day the society photographer was ill. Woolley adds, "[T]he Sun Carnival balls and parties were the biggest social events of

Three generations of Leavells are present as Pete Leavell and his artist wife, Lindsay, join the group with a basket of treats. (Courtesy, Pat Evans)

all for El Paso high socialites. Our publisher, Dorrance Roderick, was one of them. I'm still amazed I got the assignment. Anyway he had taught me the rudiments of the big good 'speed graphic' camera . . . the one you see in all the movies from the 30s and 40s."

Woolley set out in his second-hand 1951 Plymouth to the Leavells' new mansion on Red Rock Canyon Road with his hands sweating so he could hardly hold the steering wheel.

> It was a huge house to my Davis Mountain eyes. It was full of beautiful objects, paintings, sculptures, books, lovely furniture. I don't think I would have been more impressed if I had walked into Buckingham Palace. "Mr. Leavell met me at the door. He was gruff and serious-looking, but friendly. He was a good friend of Mr. Roderick, which exacerbated my already nervous and hung-over condition. He must have wondered why the *Times* had sent a kid to do the job, but he didn't let on. He introduced me to Mary Lee and then disappeared.

Woolley remembers that she was very pretty, very sweet and didn't seem at all alarmed that her big moment was being handled by a near-amateur. Woolley recalls, "Naturally, Mary Lee was depending on me to tell what kind of pictures, her dress, her poses, and all the other stuff I was supposed to know. I didn't have the haziest notion. The few photos I had taken were of murder victims, a few football games, and zero society stuff."

Woolley just made it up as he went along. He had her put on an evening gown and light a candle or something. He had her put on a sporty outfit and get out of the little three-wheeled Italian car she had, not much bigger that a golf cart. Woolley says, "I was in wonder at the toys of the rich but tried to pretend I'd seen all this stuff until I was jaded. There were a number of other pictures but I was in a daze and can't remember them."

In between clothing changes, Woolley browsed through the study and living room. He says he'll never forget that on the coffee table was a limited, special-saddle-blanket edition of Tom Lea's *The King Ranch*. Poor young Woolley spent most of the morning there with queasy visions of the film all coming out blank, knowing that he would be ostracized by the Leavells and fired by Roderick and live in disgrace forever.

After he had photographed Mary Lee in every costume and pose he could imagine, he drove as fast as he could to the *Times* to develop the film. To his astonishment (which has lasted to this day), the shots were great, crisp and clear and Mary Lee looked lovely in all of them. He says, "The city editor was pleased, Mr. Roderick was delighted. They ran about ten of them in the Sunday newspaper. We had an entire page on Mary Lee and parts of other pages as well. I took a full breath for the first time in days. Looking back on it after all these years, I think that may be the day I became a pro."

What a compliment to the Leavells, the *Times*, and Brian Woolley himself. He has now worked for the *Dallas Morning News* for a long time but is a classic southwestern novelist of such powerful books as *Some Sweet Day*.

It has already been related how the Leavells' daughter Mary Lee came right out of Stanford, married and, with her husband, heroically took over the huge J M Ranch. They handled it like old pros. Later Charles organized an El Paso corporation under the name C.K. Enterprises—it stands for "Charlie's Kids." Mary Lee is a big part of this. She continues with her book clubs in El Paso and world travels, as well as taking a part in planning and development of the 4UR while counseling and visiting her own children. She is a good and avid golfer.

The Leavells' children and grandchildren intend to carry on the traditions of beauty and dignity at the 4UR, so it is time for Charles and Shirley to speak their feelings about them. Charles says, with that familiar twinkle in his powerful blue eyes, "We were married in 1938. Mary Lee was born eleven months later, we were in kind of a hurry on this thing. I learned to changed diapers right off and we started raising a family. I built a tiny little house especially for Mary Lee.

Shirley interjects, "Well, Mary Lee grew up, of course, went to grammar school and became quite a horse girl at that early stage. She had two horses she loved very much and she rode in shows, English style, and won many ribbons and cups between the ages of eight and thirteen. She spent two years at El Paso High School and then went off to Shipley School."

Charles says proudly, "She was a vigorous girl, very cute and athletic, I tell you. Mary Lee was a very strong girl and still is as a woman. She was perfectly beautiful. When she was all of fifteen, she had some friends around that we justifiably felt would cause her great grief in the future.

"So, Shirley said to her, 'Little girl, you're going to move to a girl's school in the East.' It was real soon after that when Shirley and Mary Lee took a trip back there to look at three or four schools. She chose Shipley in Bryn Mawr, Pennsylvania. It's right across the street from Bryn Mawr College. She spent three years there before going to Stanford. When she would come home in the summertime, we took grand tours of Europe and other parts of the world. Sometimes, we'd take her with us to the Wind River Mountains of Wyoming, way back in the Indian Pass, to Horseshoe Lake, Leavell Lake and over to the Golden Trout Lakes. As you know, Tom and Sarah Lea went with us.

We'd ride, and ride, and ride. I couldn't teach her how to ride, she already knew that, but I did teach her to fish. She fell in love with one of the wranglers. A cowboy guide named Courtney Skinner. He was seventeen and she was sixteen. Young Pete was with us through all this, but we'll get to him later. That year we were coming home from Wind River and she was in the back of the car. I looked in the mirror and she was weeping her eyes out. I'll never forget it. This beautiful girl was sixteen or seventeen years old. We stopped to get a cup of coffee and Tom Lea made a classical remark about the human heart and the beauty of nature and love and everything else. I still remember the deep feeling, but not the words.

"The year she graduated from Shipley we took her and Pete, the whole family, on a grand tour of Europe. We crossed on the Andria Doria. It's the ship that was struck and sunk on the way back to New York. We had a German ex-Nazi tank-driver for a chauffeur that we picked up in Austria. He would drive up in front of our hotel in this magnificent Mercedes Benz limousine, doff his cap to Shirley, salute me and say, 'Now, children, haaas everyone gone to the toilet? Ve hab a very long trip.'

"His name was Albert Deam, a tough ex-Nazi through and through, but did what we needed to perfection. Oh, it was a grand tour for sure. We started with Spain and hit Austria, Italy, Switzerland, Germany and of course Paris, France. We went to the operas, to the ballets, to all the great museums. We saw the beautiful lakes. They had a hard time keeping me from fishing, I can tell you.

"By that time, Mary Lee had been accepted at Stanford. Her major was psychology She had a lot of fine boyfriends. I liked them, but the only one that 'took' was David Winston. His family came from Houston—prominent and wealthy people. David and Mary Lee were married their junior year.

"David studied just enough to pass. He was a wild, rough kid; I sure liked him. He was handsome and wild and only wanted to be a rancher. Well, as we know, already, he got his wish in that great mountain valley of the J M Ranch. David was a great hunter, but the thing I remember was Mary Lee getting a big buck antelope. He'd decided to run by us and crawl under the fence. As he ran his belly wasn't but six inches above the ground and that daughter of mine raised her gun and bang, shot him right through the heart and rolled him. It was the prettiest thing I'd ever seen—if you're a hunter. If you hate to kill things, like I do now, well, it hurts.

"They were a wild couple, I tell you. They were so cute. They had the tiniest little airplane and they both learned to fly it. It was the dinkiest little thing you ever saw. One engine. They would take their first baby Matthew and stuff him in the glove compartment right in the back of their seats, strap him in and away they'd go. They soon had two boys, Matthew and Freddy.

"Anyway, I've already told how they both finally wanted to go their own way after the J M Ranch years and a little about her son Freddy's wreck, but not all.

"He flipped the pickup on an icy road and became a paraplegic at the age of fifteen. We all tore up to Colorado Springs as fast as we could and got him into Craig Institute—a rehabilitation hospital. Then Mary Lee moved up there to Denver with him, and helped save his entire being. Today he is a magnificent man in a wheelchair. He's running his other grandfather's big estate, running his cousin's estate, has his own money with a management staff. He plays basketball, races in his wheelchair, hunts and fishes beautifully. He and his friend are the only people we allow to hunt up here at the 4UR. And they are damned good at it, let me tell you. [Mary Lee's] older son, Matthew, is a Doctor of Science and a fine professor. So Mary Lee is very proud of them.

"She built a beautiful house. She's financially well off, she has a good crowd of devoted friends. She has her own horses. She broke her leg badly skiing one time, so she quit that, but she dances beautifully, plays golf competitively, and she's made her own money in stocks and real estate. Very bright gutsy lady, let me tell you. The whole family is proud of her and we love her very much. She can hold her own in any society, whether it's plain old folks, or the most elegant people of New York or Paris. She's something."

Charles thinks quietly back a few moments now about Mary Lee and then begins again, just as fondly, telling about their son, Pete. In his blue eyes, shifting mental images move inward and then back out, as he says, "As a little boy we called him Black Pete—he hated to bathe.

On the trips to the Wind River Mountains he dressed like a cowboy and would never take a bath. He never was a great problem to us. He didn't take to horses and such as did Mary Lee, but he was jovial and has always been a lot of fun. I sure remember one thing he did while in grammar school.

"He entered the soapbox derby. All the neighbor kids helped him build his soap box. They built it so big we couldn't get it out of the basement. Had to take it apart and reassemble the wooden machine. Of course it looked great, but had too much air resistance. I didn't get to take him to Akron for the finals but it was fun and a great experience for all of us. After grammar school, he went to El Paso High awhile, but he was so smart he was getting bored. So Pete and I went East and looked at several schools. He chose Andover in Andover, Massachusetts.

Shirley speaks now, her entire lovely countenance smiling. "He appeared to be the nicest little boy at Andover. The truth is, he was always in trouble, doing some form of practical joke or other. Charles and I were up there for 'parents' weekend' at the school. We flew into Boston and drove up to Andover to meet with the headmaster and he said, 'I must tell you, Mr. and Mrs. Leavell, that your son has been "campused" this weekend. He can't leave his room.' We stayed on through parents' weekend with all the others asking, 'Which one is your son—' and we'd say, 'Well, he'll be here shortly.' I can't even remember now what Pete had done."

Charles can't wait to add, "So Pete graduates from Andover and goes to Stanford and hog-wild at the same time. He said to us, 'I've already had all that. I already know all they're trying to teach me. I learned it at Andover.' So our son, Pete, went to Stanford University and didn't go to school.

"Then along came a letter suggesting that Pete miss a year of school until he could decide what he wanted to do with his life. That's when Mary Lee and David were at the J M. So we sent him up there in the middle of hard freezing winter. I took him out and said, 'Son, you see that big yellow machine there—' Pete said he thought it was a tractor. I said, 'Yes, it's a D 8 tractor. It cost me a lot of money and it must be cared for better than a champion thoroughbred racehorse. You grease it at all times and you pour diesel fuel in it and you get up there on that iron seat and put the blade down and you build irrigation ditches and you improve all our roads with it.' He didn't know it but I was keeping a keen eye on him. He'd been up there three or four months and I drove out in a jeep in late February with some binoculars to watch him working. I was shivering in the jeep with the heater on watching him scratch at that frozen roadbed with the blade of that huge D8. He looked at his watch. I looked at mine. It was 5 o'clock. So he lifted the blade and started for home over a bridge that wouldn't have held up a Shetland pony. I was honking the horn and yelling for him to stop before he got killed. He couldn't hear me over the noise of the tractor and the freezing wind. That tractor went straight across the bridge and came out safely on the other side. I turned and beat him back to the house. He came in shivering, walking so stiff you expected sheets of ice to fall off him. I told him to get a hot shower and come on back. He did, gladly. I gave him a beer and I had a little whiskey. He said, 'Dad, my butt is sore as a boil from sitting on that seat. I'll do anything to get back into Stanford. I will use my brains instead of my ass from now on and I swear I'll graduate.'

"He majored in history. He knows the history of the world and graduated in the top third of his class at Stanford. By this time the Vietnam War was going on. He enlisted. He said, 'I'm not going to allow my friends and this country to have a war without me participating.'

"He went off to boot camp in the Army. Then he went to officer candidate school and became a lieutenant. Thank God he was shipped to Germany. When I think back I know he would have been killed in 'Nam. He spent four years in Germany, was an artillery man and his uncle, General Polk, gave him responsibility of the field, the use of atomic weapons, everything. Now a captain, Pete had responsibility far beyond his rank.

"He had married a lovely girl before he left and they spent four years there during the atomic standoff with Russia. When he came back he tried to work for me, but it wasn't his nature. I had set up a good trust for him, well into seven figures. A man named Claude duBock talked him into financing a documentary film in Hollywood about the

Grand Prix. They spent a whole year in Europe and made the Grand Prix film in summer season. In the meantime, he and his wife had a daughter named Jennifer. They came back and moved to Hollywood. They became a little bit on the hippie side. His wife was a second daughter of the John McGuire family—of great ranching ancestry in Texas. At first, we had a lot of fun with them. Before long we weren't getting along with them very well and they weren't getting along with each other.

"But I can't forget one trip when they came to visit us. I found Pete on a couch laughing himself silly. He said, 'You know what your five-year-old granddaughter just said as you approached the front door? She said, "Papa, is this the house where I can't say shit or pick my nose?"'

"They stayed on in California and got a divorce. Pete hung in there and made another movie and a musical that closed in San Francisco. The people he was dealing with were interested only in the wages they could draw and the money they could rip off from the budget. They cleaned out Pete's trust like emptying a vacuum cleaner."

Charles Leavell didn't know it, but he was only relating what had happened to thousands of enthusiastic greenhorns with money in Hollywood. It's the Hollywood failures who get to them almost every time. These losers have perfected this one talent only and lie in wait to ambush their innocent prey like crippled, but still deadly, lions. But out of the worst came the best.

In California, Pete met Lindsay, an artist and a wise and dedicated woman. She began to straighten Pete out from his loss of money and faith in the human beast. They got married and left for El Paso awhile later.

Pete told Charles straight on, as he had been brought up to do, "Dad, I want to move to Boulder, Colorado. It's where my wife wants to go as well. We need to be away from you all and do our own things."

Charles explains, "Well, we conceived an idea for an entire complex right there in my office. We designed all the infrastructure, all the buildings, and all the economics of the thing. Then we bought some land up there and Pete went to work."

At any rate, Charles loaned Pete $2 million he asked for but also gave him a million in cash and turned the project over to him. To say Pete did a splendid job is an understatement. He signed for all the loans—the whole thing. He developed over a hundred homes, a total development job, an entire complex, a shopping center, parks, a top-notch suburban development.

There is no way Charles and Shirley can verbally express how proud they are of Pete, for this was only the beginning of successes for their son and his family—Pete and Lindsay had a son named Brian during this period.

In the mid-1980s, Pete's Boulder, Colorado, company, Leavell Management Group, bought a Renaissance Faire outside of Chicago. The Faire was on the verge of collapse, but within two years they were able to turn it around and help it start making good money.

These Faires give the public of every age great colorful and historical entertainment. They have everything from jousting matches, with knights in armor on the finest steeds, to blacksmithing and sewing exhibitions. They present all the old dances, songs and music, with the participants in absolutely authentic costumes of the Renaissance period. There are storytellers and playlets of many forms for both children and adults. It is ancient pageantry of the highest order of pure entertainment, with the audience participating where suitable. Color, color, color and action are the essence of the Faires.

After their first success in Chicago, it became apparent to Pete that this could develop into one of those extremely rare businesses that is profitable and does a lot of good for an eager public at the same time. They merged Renaissance Faire into a public company.

As Renaissance Entertainment Corporation (REC), their next move was to acquire Faires in San Francisco and Los Angeles, which were not only the original ones, but also two of the largest in the United States. These two Faires had lost millions over the past two years. Once REC inserted its highly skilled and dedicated management team, they turned a profit the very first year. At this point REC enlisted the help of a New York investment banker, and in January 1995 REC had a public stock offering that raised $3.5 million. Attached warrants, once exercised, would raise an additional $7 million. In Pete's words, "We were smokin'!"

In the summer of 1995, REC acquired the New York Renaissance Faire and property to build a Faire from scratch in Virginia, just south of Washington, D.C. This was to open in May 1996. On September 1, 1995, REC met the criteria to be listed on the NASDAQ stock exchange; as of November 7, 1995, this meant the company had a market capitalization of $34.5 million. The stock was trading at $12.50 a share. All this from an initial actual investment of $400,000!

In April 1995, Pete Leavell resigned as chief executive officer and moved upstairs as chairman of the board. This was possible because he knew he had a fine team of people and enough money in the bank so that they could easily run REC without him on a minute-to-minute basis. Two things are extremely important here. Like his father, Pete has that strange sixth sense that guides his to acquire the best people for his endeavors. Next, his executive shift meant he could now give more time to his main love, as all the Leavell family do: the 4UR Guest Ranch.

At the request of his father, Pete took over management of the 4UR for the family in 1988. At the time, Joe and Katus Walton had been general managers for seventeen years. Pete explains, "They were ready to retire after making indelible marks on the ranch." It was at this time that Rock and Kristen Swenson were hired. Pete says they have done a marvelous job from the first day.

Pete and the Swensons set out with a ten-year plan to make the 4UR one of the finest family resorts of its kind. "It's a labor of love," he states with much feeling. They have constantly upgraded everything from the water systems, cabins, the lodge, the laundry and the kitchen to the heart of it all, the cold mountain fishing creeks. They have sophisticated computer accounting and reservation systems but still keep the feeling of the historically significant past. Pete says, "We've been right on schedule and soon I will have to come up with another ten-year plan."

Charles and Shirley Leavell's son surely echoes the sentiment of their own hearts and soul-attachment to this special place as Pete says with total openness, "I love the place more than any physical location in my life and when I am there I feel close to my soul's center. I have ridden horseback, hiked and driven all over that country and know it better than my own neighborhood in Boulder. I feel God has given me the responsibility via my father of stewardship of that place and I take it very seriously. I intend to pass on its importance—this same sense of responsibility—to my son."

All of us who love the spiritual magic of the lovely earth, with its vegetation, water and wildlife, can take heartfelt consolation from Pete Leavell's words. The 4UR aura moves on in the best of caring hands. If only there were millions of others like these giving people to care for the giving land!

CHAPTER 25

The Big Bow Knot and the Forever Time

Charles H. Leavell recently gave the 4UR to his children, completely confident of its proper perpetuation. They all have their own cabins there. Charles and Shirley's cabin is nearest to the lodge and its gourmet dining. This dwelling faces out on the large, heavily grassed park-like area looking across the meadows to Goose Creek and the mountains on both sides. Their cabin is elegantly decorated with art objects from their many travels, such as Masai spears and masks, English prints and art work from Creede artists.

It took Charles twenty-six years to acquire his dream, and he has now owned it for exactly twenty-six years. During that period the Leavells have enjoyed all the changing seasons and the changing children that play there year after year—many until they are grown and have their own offspring. They have hunted, fished, worked, planned and visited with all the people. They have enjoyed the wild plants and wildlife of the waters, mountains, forests and meadows with increasing pleasure and love. Twenty-six years of agonizing, twenty-six years of constantly growing joy—a fair tradeoff; a 4UR formula of dedication seldom rivaled.

Winter is coming; the summer of guests (the 4UR family) is at an end. The last ones are loading their various conveyances with luggage, fishing tackle and children. Most will have stories to tell all winter—and, in fact, for the rest of their lives—about the big one that didn't get away. Memories will be taken with them across America and much of the world; memories of sudden thunder showers, nurturing everything as the sun breaks through in tilted golden highways of the sky to turn the grass greener and the mountains bluer. The animals, tame and wild, are fatter, sleeker. The summer visitors will carry the camaraderie of the 4UR family with them in peaceful thoughts that help their bodies and souls survive the trials of a high-tech, motorized, computerized, overpopulated world The guests spread these good feelings to countless others until they can return to their second home to be revitalized once again.

Then the leaves of the aspen turn quivering gold across the mighty Rockies of the entire 4UR realm. The bears feed ceaselessly on berries to fatten even more for the long months of sleep in caves—some of which have been inhabited for hundreds, probably thousands of years by their genetic ancestors, to a time before the Utes. The fall air is briskly cool and clean. It is the "quiet time" down at Taos Pueblo when all the people, plants and creatures of the earth rest and appreciate.

Here on the 4UR, it is naturally the same. Except, like the bears, Rock and Kristen and a few other hands prepare the cabins and outbuildings for winter vacancy. Only the spirits of the departed, both living and deceased, remain on guard, until the 4UR family returns the next summer.

And now the creeks start freezing on the edges at night, in warning to all living things to heed and prepare for the great white blanket to be thrown across all the land. The squirrels and chipmunks work in a frenzy, gathering nuts and seeds for sustenance during the white-blanket time. The bugling breeding call of the elk creates a mystical symphony across the entire range.

The morning ice moves nearer the center of the creeks each morning, and the coyotes hunt harder in the longer nights. Then the grey-dark, purple-black cloud gods come swooping in, and with billions of hexagonal flakes drape the mighty white blanket over everything, covering all from the highest peaks down to the valleys. Only the sheer bluffs of Wagon Wheel Gap and other rocky extrusions have immunity from the blanket, but they pay for this blessing by being fully exposed to frigid blasts of Rocky Mountain air that crack and erode them with infinite slowness. Nothing fights to escape the realm's blanket, for it is

The 4UR compound from over ten thousand feet elevation in the southern Colorado Rockies. (Courtesy of Pat Evans)

a blessing of bountiful liquid for the spring sun to alter into the sustenance of everything living.

For Rock and Kristen Swenson, the work, chores and total property management continues unabated: feeding horses, keeping the roads open with tractors and keeping the snow off rooftops, maintaining the machinery, and on and on. Kristen does the buying to restock the summer gift shop and complete all sorts of interior renovations. Rock says, "Surprisingly to many, the winter never seems long enough."

The elk come down from the deeply blanketed high country for fall breeding and easier winter feeding. Where one hot spring is piped underground to a tank, it warms the earth and often melts the snow; some of the elk bed down there at night for a warming rest, in full, fearless view of the Swensons.

257

In late November 1995, Rock spotted something moving about in Goose Creek that was not a fish. On examination, he found that an elk calf had broken through the ice and was in danger of drowning and certainly would freeze to death shortly. The mother elk was helpless to save her baby. Rock and Kristen jumped right in the freezing water and with great difficulty pulled the frightened, wildly kicking calf to safety. They put it in a horse stall with dry straw and dried it with blankets. While Kristen desperately tried massaging the circulation back into its dangerously shaking body, Rock ran and phoned the local game warden, officer Woodward. He came as fast as the icy roads would allow and gave the calf two intravenous feedings. All three kept up the massaging till the calf stood up and watered and fed from buckets. They had saved it. There was nothing left to do then but turn it loose to find its mother. Nature. Just part of an evening's work in the 4UR realm.

Of course, as always, the clouds begin to thin, the wind becomes less freezing, and the sun daily spends more time in the valley. The bears start to awaken, drowsy and sluggish. Then the blanket of white turns softer in the daytime and icier on top at night. Millions, billions of tiny streams start seeking lower creases of earth to move down into. The trickles grow bigger as the snow drifts shrink. The creeks and rivers rise; the grass pushes its tender shoots up out of the thawing earth; the aspens bud and turn clean spring green; the birds return from their winter jaunts; the coyote howls in joy as the ice melts on the streams, then disappears. The thinning white blanket's torrents pour in from all sides and the streams often go out of their banks as if trying to escape from the plunging cold roar of the mainstream. The fish feed. Everything feeds. It is spring and the 4UR family will soon be preparing to come in their allotted summer times, to meet and renew once more. And again the coyote howls in joy and is answered. Kristen and Rock and all the ranch-hands are ready for the new season, just as all the earthly things are ready for theirs.

Each part of the world has its own unique fragrance. The millions of years of falling leaves, trees, grass and vegetation of all kinds found in the dirt, the earth around and in the seams of the rocks, the people, the domesticated and wild animals: each contributes its own private scent. From the *before time* to the *now time*, that heady melange becomes the true essence of a place.

Besides the singular fragrance of a place, there are the individual sounds, the degrees and distances of human voices. The wind caresses the trees, the bushes, the grass and all the special formations of the mountains and their forests, and all harmonize to create one distinguished voice that penetrates to the deepest recesses of human and animal awareness. It becomes a permanent part of them wherever they are on earth. It tells the memory cells to give instant visions and smells of past pains and pleasures, of that particular place, that are passed on to the *forever time*. The fragrances and sounds of the 4UR most certainly continue, from hour to hour, year to year, lifetime to lifetime, from long before the Utes to the Utes to the initial homesteaders and prospectors to General William Palmer to Allan Phipps to Charles H. Leavell and the 4UR family, in the blood and breath and being of all.

Charles and Shirley's life has been a lasting love affair, between themselves, their family and their friends. It has been one of great risks and great rewards. One must hope that there are many sensitive people left to have enjoyed and appreciated them and their lasting dedication to the sacredness of the 4UR realm.

This 4UR realm is as big as the flight of a thousand eagles, but it can feel small enough to be extremely personal as well. There once was a member of the 4UR family named Dick Shaw. Charles says, "He was one of the finest hunters, fisherman, and all-around fine fellows I've known in my life. He was president of the W R Weaver Company, makers of the great hunting guns and scopes."

Shaw's wife had died several years before, at their home in El Paso, and for years Dick had not dated anyone. Then he hesitantly told his dear friend Charles about a fine lady he had met at a club in Denver. He was shy about even asking her out to dinner. Charles told him strongly, "Now, Dick, you take that lady out to the finest place in Denver, and when you bring her home, be polite, but don't go in or even try to kiss her. Let things work naturally. From what you tell me about her, people like Toni are scarce and priceless."

Dick Shaw fearfully followed Charles's instructions. It worked. Dick and Toni soon married and happily spent many years at the 4UR together—and with the Leavells. Then Shaw got incurable cancer and asked for Charles to come visit him. Charles immediately went to see Shaw. They sat and reminisced about all the wonderful adventures they had shared on the 4UR. Shaw said, "That's the greatest place I've ever been, and my wife, Toni, is going to bury me there." Just like that. He told Charles that he had something special he wanted to give him— something to reward him for all the years of great friendship and favors.

Shaw said, "You've shared your ranch and who you are with me. You helped me court Toni, and I want you to have something you'll remember all your life."

Charles relates, "He walked into another room and came back smiling proudly and handed me the most gorgeous shotgun you've ever seen. It was handmade at his company by their greatest craftsman. I was touched deeply. He gave me the gun and a month later he died. We buried him next to a great boulder in the aspen grove on the south side of the valley so he could look out over a meadow towards his beloved Goose Creek as long as there is a world. His wife Toni bought and occupies the remodeled train depot just across the Rio Grande at Wagon Wheel Gap, right where the Goose Creek joins the great river for its long journey to the Gulf of Mexico. She can see the aspens where her husband rests amidst profound beauty."

Ed Cook, son of Vernon Cook, has been coming to the 4UR since he was a tiny boy fifty years ago, accompanying his father and family. His favorite story—everyone of the 4UR family has one—concerns two good-looking twin daughters, Joan and Jean, of the Neal Johnson family of Amarillo, Texas, and their run-in with F. T. Firstman. Firstman felt so at home at the 4UR that he'd have his personal furniture moved back and forth each summer. He really proved his deep love for the 4UR, as Vernon Cook would later learn.

Firstman claimed to know every inch of Goose Creek and the Rio Grande. More than anyone else, he prided himself on knowing where all the big fish were. All fish were wild then; nothing was stocked. The Johnson twins would fish, but also enjoyed sunning themselves and simply exploring, enjoying the smells and noises of the eternally moving water. One of them, Joan, is still coming back as Mrs. Jack Shelton.

One day, the twins came in for the noon meal with a huge trout. Firstman immediately recognized the trout and described the hole it came from. He kept on trying to get them to agree that he was right. The twins never gave him a direct answer because, in fact, they had found the fish trapped in an eddy and had clubbed it to death with sticks. F. T. Firstman was mad at their catching his prize fish and he died mad. Such is the obsession a Goose Creek trout can create!

Izaak Walton, who was also a most dear lover, and a frequent practiser, of the art of angling, said this when writing about Sir Henry Walton in his world-famous book, *The Compleat Angler:* "[I]t was an employment of idle time, which was then not idly spent . . . a rest to the mind, a cheer booster of his spirits, a diverter of sadness, a calmer of unquiet thoughts, a moderator of passions, a procurer of contentedness; and that it begot habits of peace and patience in those that professed and practiced it."

Vernon Cook, along with Dick Shaw, cherished this quotation, which for them utterly describes the 4UR, for their whole lives. Charles tells of Vernon saying, just before his impending demise, "I'm getting down there with those trout, by golly, and I'm gonna find out why they're so damned smart." Ed and the Cook family took Vernon Cook's ashes way up to his favorite fishing place, Station 15, and lovingly scattered him among his watery friends.

A habitual guest of the 4UR, Bradford Young of San Francisco, wrote music and lyrics for a song about it. Here are a sample of the lyrics. The music will come to you when you listen to Goose Creek's eternal voice.

> There's a moose on the loose, up on little Goose Creek,
> So you say what's the use, he'll be there for a week,
> Just a munchin' on moss with his ugly old beak!
> It's a Rocky Mountain Fairy Tale . . . Rocky Mountain Fairy Tale.
>
> What's the allure there? The air is pure there—
> It's lucky you're there, and not in Detroit—
> Wish for a fish—try a fly on fourteen,
> And if one is amiss, try a hole in between;
> here's a trout thereabout that's so big its obscene.
> It's a Rocky Mountain Fairy Tale . . .

Blue, clouds are few, not a chance it will rain;
Then a mist hides the view and it's winter again—
Later on stars are bright—in the late summer night.
It's a Rocky Mountain Fairy Tale.

And so it is, albeit a very true one. Cook and Shaw watch over it for the returning 4UR family, and the Leavells watch over its rarefied beauty and uncommon dignity for us all.

Now Mr. Charles H. Leavell, of El Paso, Texas, and the world, says simply, "So, there's two old pros buried up there on that ranch. They both said 'Bury me in that special place. That chosen place.' and that has to be the immeasurable meaning of love."

Sources

1: This Chosen Place

Leavell, Charles H. Taped interview with author. January 1995.

2: The Valley of Sudden Death and Bountiful Giving

1. Randles, Anthony (Slim). Interview with Alden Narajano. January 1995.
2. Ellis, Richard M. (Fort Lewis College, Durango, Colorado). Letter to author regarding the Utes. December 19, 1994.
3. Leavell, Charles H. Tapes and interview with author. August 1995.

3: Roots of Home and Guts Aplenty

1. Coleville, Rose Marie. *San Luis Historian* 26, no. 4 (1994).
2. Kessler, Ron, and Phil Carson. "The Journals of Don Diego de Vargas." *Spirit* (fall/winter 1994–1995), pp. 8–10.
3. "Journal of the Vargas Expedition." *Colorado* magazine, May 16, 1939, pp. 81–90; "In the Footsteps of Don Diego Vargas; Monday, July 12, 1694 report." *San Luis Valley Historian*, pp. 40–43.

4: Back When the Horses Came Running

1 Leavell, Charles and Shirley. Tape recordings. 1994–1995.
2. Lynde, Bill. *The Leavell Story*. El Paso, Tex.: Guynes Printing,1983.

5: THE LAUNCHING OF A GIANT

1. Milne, Lorus J., and Margery Milne. *The Mountain.* Life-Nature Library series. New York: Time-Life Books, 1980.
2. *Bufton's Universal Encyclopedia.* Kansas City, Kan.: n.p., 1921.
3. Armstrong, Ruth. *New Mexico—From Arrowhead to Atom.* New York: A. S. Barnes, 1969.
4. Richmond, Patricia Joy. *Trail to Disaster,* with a foreword by Mary Lee Spence. Niwot, Colo.: University Press of Colorado/Colorado Historical Society, 1990.

6: CONSUMMATE COURAGE, WHITE DEATH AND MAKING HISTORY

1. Lynde, Bill. *The Leavell Story.* El Paso, Tex.: Guynes Printing, 1983.
2. Leavell, Charles and Shirley. Taped interviews. 1994–1995.
3. Armstrong, Ruth. *The Chases of Cimarron.* Albuquerque, N.M.: New Mexico Stockman, 1981.
4. Carter, Harvey L. *Zebulon Montgomery Pike, Pathfinder and Patriot.* Colorado Springs, Colo.: Denton Printing, 1956.

7: THE DAYS OF SURGING POWER AND OVERCOMING

1. *White Sands Missile Ranger.* 16 September 1977.
2. Bason, Jimmy R. Interviews. April 1996.
3. Leavell, Charles H. Interview with author. April 1996.

8: GENERAL PALMER: CIVIL WAR HERO AND OWNER OF THE 4UR

1. Wilcox, Rhoda Davis. *The Man on the Iron Horse.* Colorado Springs, Colo.: Martin, 1959 (reprint 1990).
2. Martin Associates Rotary International-El Paso. Program of events. 13 April 1995.
3. Anderson, George LaVerne. *General William J. Palmer, Man of Vision.* Colorado Springs, Colo.: Colorado College Studies, 1936.

9: THE PHIPPS CLAN OF COLORADO: DOERS AND GIVERS SUPREME

1. Hubka, Michael A. *The Phipps Brothers, Building of an Empire.* Thesis (quoting *Denver Post* archives [September 1, 1974, p. 56; March 8, 1967, pp. 56–57] and *Rocky Mountain News* archives [February 16, 1963, pp. 46–48]).
2. Juhl, Derk W. *The Phipps Family: A Denver Dynasty.* Master's thesis, University of Colorado at Denver, 8 May 1981.
3. Ducey, Audree. Final project—Graduate Students' papers—Denver History 340., University of Colorado at Denver, spring semester, 1987.
4. Noel, Thomas J. Report to author, University of Colorado at Denver, April 1996.

10: EXPLOSIONS OF ENERGY CATAPULTING THROUGH CANYONS OF DREAMS

1. Lynde, Bill. *The Leavell Story.* El Paso, Tex.: Guynes Printing,1983.
2. Leavell family. Taped interviews with author.

11: THERE IS NO NIGHT IN CREEDE, OR SO THEY ONCE SAID

1. Bennet, Edwin Lewis. *Boom Town Boy in Old Creede, Colorado.* Chicago: Sage Books, 1968.
2. Settle, Jr., William A. *Jesse James Was His Name.* Kansas City, Mo.: University of Missouri Press, 1964.
3. Horan, James D. *The Trial of Frank James for Murder.* Kansas City, Mo.: George Miller, Jr., 1898.
4. *The Gunfighter.* New York: Time-Life Books, 1974.
5. Feitz, Leland. *Creede, Colorado Boom Town.* Colorado Springs, Colo.: Little London Press, 1969.
6. Smith, Toby. *Bid Blackie.* Wayfinder Press, 1987.

12: GO WEST, YOUNG GENERAL, GO WEST TO THE COLORADO ROCKIES

1. Anderson, George LaVerne. *General William J. Palmer, Man of Vision.* Colorado Springs, Colo.: Colorado College Studies, 1936.

2. Sprague, Marshall. *Newport in the Rockies: The Life and Good Times of Colorado Springs.* Denver, Colo.: Swallow Press, 1987.

3. Fisher, John S. *A Builder of the West: The Life of William Jackson Palmer.*

13: Charles H. Leavell's Power on the Ground and in the Air

1. Leavell, Charles and Shirley. Taped interview. El Paso, Texas. 4 May 1995.

2. Lynde, Bill. *The Leavell Story.* El Paso, Tex.: Guynes Printing, 1983.

14: The Leavell Company Joins the Cold War with Hot Projects

1. Leavell, Charles H. Taped interview with author. El Paso, Texas. April 1995.

2. Lynde, Bill. *The Leavell Story.* El Paso, Tex.: Guynes Printing, 1983.

15: The Wiles of Walton and the Last Trail Winding

1. *William J. Palmer, Pathfinder.* George Foster Peabody.

2. *Rails That Climb: The Moffat Road.* R. R. Bowker/Colorado Railroad Museum, 1979.

3. Lipsey, John J., comp. *The War of the Gauges.* Colorado Springs, Colo.: private printing, 1968.

4. "Scenic Byway." *Silver Thread* [Lake City, Colorado], Summer, 1995.

5. Leavell, Charles H. Timeline from private papers.

16: Buying Up the World and Hoping for Paradise

1. Leavell, Charles and Shirley. Taped interviews. El Paso, Texas. 1994–1995.

2. Chapell, Virginia. *El Paso Times.* 28–30 May 1956.

17: THE HIGH AND THE LOW: A NECESSARY LOVE AFFAIR

1. Simmons, Virginia McConnell. *San Luis Valley, Land of the Six Armed Cross.* Boulder, Colo.: Pruett Publishing, 1928.
2. Bean, Luther E. *Land of the Blue Sky People.* Alamosa, Colo.: Ye Olde Print Shoppe, 1975.
3. *San Luis Historian* no. 2 (1995); no. 2 (1993); nos. 3 and 4 (1984).
4. Feitz, Leland. *Creede: Colorado Boom Town.* Colorado Springs, Colo.: Little London Press, 1969.
5. Vance, Beryle, and Andrew Pagett. *Jack Dempsey, Manassa Mauler.* Alamosa, Colo.: San Luis Valley Historical Society, 1976.

18: NATURAL LIVING AND SOME TRANQUILLITY

1. Leavell, Charles and Shirley. Taped interviews with author. 16 May 1995.
2. Wintz, Ed. Taped and written interviews with author. Alamosa, Colorado. February 1995.
3. Geis, Bill and Betty. Taped and written interviews with author. South Fork, Colorado. Spring 1995.

19: REPAIRING PARADISE JUST A LITTLE

1. Leavell family. Interviews with author. 16 May 1995.
2. Quotes from Hayden Survey Party, 1877. *Geologic and Geographical Atlas of Colorado and Portions of Adjacent Territory.* Washington, D.C.: Department of the Interior.
3. *Creede Candle.* 1893. Quoted in Leland Feitz, *Creede: Colorado Boom Town.* Colorado Springs, Colo.: Little London Press, 1969.

20: TIME OF REBIRTH AND FRIENDSHIPS RARE

1. Korzeb, Stanley L. "The Wagon Wheel Gap Fluorspar Mine, Mineral County, Colorado." *Mineralogical Record* 24, no. 1 (January/February 1993).
2. Publication by Institute of Texas Cultures, 1969.
3. *Silver Thread* [Lake City, Colorado]. 1995.
4. Lea, Tom. *The Southwest: It's Where I Live.* Dallas, Tex.: De Golyer Library, Southern Methodist University.

5. Craver, Rebecca, and Adair Margo, eds. *Tom Lea (An Oral History)*. El Paso, Tex.: Texas Western Press, 1995.

6. Hjerter, Kathleen G. *The Art of Tom Lea*, with an introduction by William Weber Johnson. College Station, Tex.: Texas A&M University Press, 1989.

7. Dingus, Anne. "War Paint." *Texas Monthly* (August 1994), pp. 89–93.

8. Kleberg, Mary Lewis. Draft of speech "Salute to Tom Lea." 4 May 1990.

9. Leavell, Charles H. Speech introducing the Tom Lea Hall of Honor, El Paso Historical Society. 9 November 1975.

21: SEEKING THE LOST LAKES: THE DEMOLISHING OF BURDENS

1. Leavell family. Interviews with author. 20 May 1996.

2. Russell, Keith C. *The Fly Fishingest Gentleman*. 1978.

22: THE DAILY WORKINGS: FULFILLED YEARNINGS

1. Leavell family. Interviews with author. 20 May 1996.

2. Swenson, Kristen and Rock Swenson. Interviews with author. 5 December 1994; 10 May 1996.

3. Wintz, Ed. Interviews with author. 5 March 1995.

4. Geis, Bill. Interviews with author. 6 March 1995.

5. "Julia Child." *New York Times* [Associated Press release]. Winter 1996.

23: MINES AND MOGULS: SAVIORS COME CALLING

1. Korzeb, Stanley L. "The Wagon Wheel Gap Fluorspar Mine, Mineral County, Colorado." *Mineralogical Record* 24, no. 1 (January/February 1993).

2. Emmons and Larsen. *Geological Report* [to United States Department of Mines]. 1913.

24: GENES OF HISTORY: THE FAMILY OF PERPETUATION

1. Leavell family records.

2. McCarty, Jeanne. "Shirley and Charles Leavell Awards." YWCA Anniversary publication, El Paso, Texas, 27 April 1994.

3. Leavell, Charles and Shirley. Interview with author. 4UR Ranch. Summer 1995.
4. Leavell, Pete. Interview (written and telephone) with author. Winter 1995.
5. Renaissance Faire brochure. Boulder, Colo., 1995.
6. Woolley, Brian. Interviews with author. El Paso, Texas. 26 June 1995.
7. Woolley, Brian. Letter to author. Dallas, Texas. 20 September 1995.

25: THE BIG BOW KNOT AND THE FOREVER TIME

1. Leavell, Charles and Shirley. Interviews with author. 1994–1996.
2. Swenson, Rock and Kristen. Interview with author. 10 May 1996.
3. Cook, Ed. Interviews (telephone) with author. Oklahoma City, Oklahoma. October 1995.
4. Walton, Izaak. *The Compleat Angler.*
5. Young, Bradford. Song about 4 UR. San Francisco, 1993.
6. Leavell, Pete. Interviews with author. 7 November 1995; 21 November 1995.

Index

Photo references are in bold type.

Abrahamson, Bill, 133, 141
American Fluorspar Mining Company, 236
American Institute of Steel Construction award, 96
Amethyst mine, 101, 235
Amistad Dam, 95, **96**
Anasazi, 44
Anderson, Robert O., 205
Antlers Hotel, **154**
Apaches, 113–14
Apollo program, 146
Arapahoes, 11
Atchison, Topeka, and Santa Fe Railroad, 121
Aunt Lee, **16**

"Babe," 18, **19**
Backus, Henry, 176
Baker, George, 167–68
Barber, Susan McSween, 163
Barnard, Jimmy **231**
Bason, Jimmy R., 60
Bautista de Anza, Juan. *See* de Anza, Jan Bautista
Bean, Luther, E., 177
Benton, Jessie, 48
Biedel, Mark, 176
Big Medicine, 11
"Blowhard," **13**
Boozer, Elvira, 240
Bourke, Mary "Calamity Jane," 102, **105**
Breckinridge, Thomas, 54
Brown, Darcy, 188
Buckley, Harry, 87
Buxton, Keith, 139–40

C.F.I. *See* Colorado Coal and Iron Company
"Calamity Jane." *See* Bourke, Mary "Calamity Jane"
Callier, Nix, 36
Campbell Soup Company, 88
Carnegie, Andrew, 68, 75–76
Carson, Kit, 6, 49
Carson, Phil, 27, 28–29
Cattle
   Longhorn, **170**, **197**
   rustling, 18, **20**
Champion mine, 101
Chappell, Virginia, 162, 165–66
Cheatham, Jack, **231**, **234**
Child, Julia, 229–30, 232
China, 135–40, **138**
Chino, Wendell, 161–62
Civil War, 64–74
Clint Texas Ranch, 12
Cockrell Ranch, 158
Coghlan, Pat, 162–63
Cold War, 141–47
Collins, S. B., 236
Colorado Coal and Iron Company, 121, 236
Colorado Springs, 66, 117–19, **154**
Colville, Ruth Marie, 25, 50
Comanches, 11, 27–29
The Complete Angler (Walton), 261
Cook, Vernon, 260–62
Cougar Country, 232
Cousin, Ralph, 232, **233**
Coventry, Mar, 63
Cox family, 58–60, 164–65
Creede, Colorado, 98–110, **100**

Creede, Nicholas C., 99
Crittenden, Thomas, 104
Cuerno Verde, 26–29
Cunningham, Eugene, 211
Curtis, Edward S., 77

Daugherty, Old Man, **13**
Davis, Jefferson, 72–74
Davlin, Chuck, 186–87
Dawes, Charles G., 191
de Anza, Juan Bautista, 26–29
de Ornate, Juan, 26
de Varga, Don Diego, 23–26
Deam, Albert, 247
Defense contracts, 142–47
Dempsey, Jack, 181–82
Denver, 75
Denver and Rio Grande Railroad, 112,
    120–23, 149, **150**, 151
Denver Republican, 102
Derrick, William A., 95
Diaz, Porfirio, 120–21
Diego de Varga, Don. *See* de Varga, Don
    Diego
Dighton, Sarah. *See* Lea, Sarah
Dobie, J. Frank, 212
Doheny, Edward L., 164
Dravo Corporation, 94
duBock, Claude, 250–51
Ducey, Audree, 78
Dworshak Dam, 94–95

Eisenhower, Dwight, 191
El Paso Natural Gas Company, 95–96
El Paso Times, 21
Elliot, Charles, 66
Engineering News Record, 95
Ethyl mine, 101
Evans, Gertrude, 212
Evans, William Gray, 77
Everbart, H. P., 163

Fall, Albert Bacon, 162–64
Farah, Willy, 188
Figure Two Ranch, 12, **13**, 22, 157
Firstman, F. T., 260–61
Fishing, 204–8, **207–209**, **220**, **228**
Flaming Gorge Reservoir, 96
The Fly Fishingest Gentlemen (Russell),
    219–21

Fonda, Jane, 60, 205
Ford, Charlie, 103–6
Ford, Robert "Bob," 102–7, **106**
4UR, **227–29**. *See also* Wagon Wheel
    Gap Ranch
    description, 185–89, 255–59, **256**
    fishing, 204–8, **207–209**
    history, 148–56
    management of, 222–25, 253
    repair of, 191–95
Fremont, John Charles, 47–56
Frizzell and Company, 91
Frizzell, Ralph, 91

Garret, Pat, 58
Geis, Bill, 191, 202–3
Glen Eyrie, 117–19, **118**, 152–53
Godey, Alexis, 52, 53, 55–56
Goldberg, Vicki, 204
Golfing, **16**
Goodwin, Charles, 148
Goose Creek hot springs, 148–49
Gould, Jay, 151
Graham, Clarence, 19
Greenhorn, Chief. *See* Cuerno Verde
Grundy, Sherman, 18, 167–68

Hansen, Peter, 176
Henson, Henry, 14
Herzog, 211
Holland, Jimmy, 239–40
Holland, Tom, 240–41
Homart Shopping Centers, 94
Hot Springs, Goose Creek, 148–49
Houser, John, 211
Hubbard, George, 50
Hughes, Gerald, 77
Hunter–Grisham Oil Company, 22
Hurst, Bob, 95

Indians. *See* Native Americans
Ingersoll, Ernest, 149

J M Ranch, 166–74, **167**, **171**, **173**
James, Jesse, 103–4
Jonesville Lock and Dam, 95

Kansas Pacific Railroad, 114, 119
Kerber, Charles, 176
Kern, Ben, 54

Kessler, Ron, 27, 28–29
Khartoum, 131–35
Kleberg, Mary Lewis, 216

Ladder Ranch, 60, 205
Lakes, 217–21
Land of the Blue Sky People, (Bean), 177
Lea, Joe, 86, 167
Lea, Sarah, 212, 246
Lea, Tom, 86, 171, 194, 208–, **210**, 246
Lea, Tom, Sr., 211–12
Leavell, Brian, **243**, 252
Leavell, Charles H., **196**, **210**, **218**, **242**.
   See also Leavell Company
  ancestors, 238–41
  as a baby, **15**
  as a young man, **31**
  at Stanford, 30–32
  childhood, 12–22, **15**, **16**, **17**, **22**, **34**
  early career, 30, 37–41. *See also* Leavell
    Company
  grandparents, **73**, 74, 238
  illness, 14, 63
  international travel,, 125–39, **127**,
    138
  purchase of 4UR, 185–89, **187**
  ranch ownership, 157–74, **172**
Leavell, Charles H. Sr., **19**, **22**
Leavell, John, 214
Leavell, John H., 239
Leavell, Lindsey, **243**, **244**, 251–52
Leavell, Mary Lee, 40, 243–48, **243**
  education, 85
  ranch life, 158–59, 167–70, 246
  travel, 139
Leavell, Pete, 40, 194, **243**, **244**
  childhood, 248–54
  education, 85
  travel, 139
Leavell, Shirley, **196**, **226**, **242**, **243**
  ancestors, 241
  childhood, 33–37, **35**
Leavell Company, 39, 57–63. *See also*
  Leavell Development Company
  Cold War, 141–47
  construction projects, 86–97
  international business, 125–40
Leavell Development Company, 91. *See*
  also Leavell Company
Leavell family, **243**

Leavell Management Group, 252
Lee, Aunt, **16**
Leftwich, J. D., 157
Leopold, King, 191
Liberia, 129–30, 141
Little Medicine, 11
Longhorn Breeder's Association, 158
Longhorn cattle, **170**, **197**
Los Alamos, 62
"Lost Lakes,", 217–21
Lowman, Al, 212
Luckman, Charles, 89
Lynde, Bill, 87

Masterson, William Barclay "Bat," 102,
  103
Matkin, George, 39
Maxwell, Lucien, 48
Mayfield, Davis, 40
Mayfield, Tom, 40
McCarthy, Carmac, 211
McClelland, Job C., 149
McKee, Robert C., 92–94
Mead, Albert, 148
Mears, Otto, 177
Meeler, Mrs. 168–70
Melchior, Lawrence, 205
Mellen, Queen, 115
Mesa Verde, 42–44
Mexican National Railway, 121, 122
Miller, Frank, 137
Mining, 98–102, 235–37, **237**
Moffat, David, 99
Moor, Lee
Moore, Joe, **138**
Morgan, J. P., 76–77
Mount Blanca, 10

Narajano, Alden, 11
NASA, 146
Native Americans
  Anasazi, 44
  Apaches, 113–14
  Arapahoes, 11
  Comanches, 11, 27–29
  Plains, 44–47
  Utes, 9–11, 23–25
Neiderhardt, George, 176
New Guinea, 139–40
Nicollet, John, 48

Norton, John, 212

O'Kelly, Ed, 107–108, **108**
"Old George," 114, 117
Orndorf, Lee, 38

Packer, Alferd G., 44–45
Pagosa Springs, 11
Palmer, William Jackson, 8, 149–56, **152**
    Civil War, 64–74, **65**, **67**
    railroads, 111–24
Paradise Island, **88**, **89**
Park Central, 91, **92**
Park Plaza Apartments, **90**
Peterson Lumber Company, 37–38
Phipps family, 8, 75–84
Pike, Zebulon Montgomery, 45–47
Pikes Peak, 45
Pinkerton, John, **243**
Pinkerton, Kelly, **243**
Pinkerton, Mary Lee. *See* Leavell, Mary
    Lee
Pinkerton, Tip, **243**
Plains Indians, 44–47
"Poker" Alice. *See* Tubbs, "Poker" Alice
Polk, James H., 161
Ponder, Dan, 62
Porter, Marvin, 158
"Pregnant Gypsy," 87
Preuss, Charles, 48, 49, 52

Railroad development,, 112–24
Ranch ownership, 157–74, **172**
Rancho Felix, 157
Rancho Seco Nuclear Power Plant, 125
REC. *See* Renaissance Entertainment
    Corporation
Renaissance Entertainment Corporation
    (REC), 252–53
Richmond, Patricia Joy, 50, 52, 56
Riebe, N. J., 86
Rio Grande Industries, 83, **84**, 87, 91
Rio Grande River, 4, 6, 99
Rio Grande Western, 123–24, 151
River Bend Ranch, 158–59
Roberts, Mary Jane, 57
Robidoux, Antoine, 51
Robinson, John, 46
Roddenberry, Gene, 211
Roosevelt, Theodore, 33–34, 153

Rourke Cattle Company, 158
Russell, Grace, 229
Russell, Keith C., 219
Ryan, Thomas Fortune, 160, 162, 165

Salt Wars, 12–14
San Juan Pueblo Indians, 25
San Luis Valley, 9–11, 175–83, **178–80**,
    181
Sanders, Bill, 95
Scott, Bill, 136–37, **138**, 142
Scott, Thomas A., 68
Sharp, Arthur, 7, 197–98
Shaw, Dick, 259–62
Shultz, James, 176
Simmons, Virginia McConnell, 176–77
Sinclair, h. F., 164
Skinner, Clem, 215
Skinner, Courtney, 247
Smith, George L., 99
Smith, Jefferson Randolph "Soapy," 106,
    106, 109
"Soapy" Smith. *See* Smith, Jefferson
    Randolph "Soapy"
Sonnichsen, C. L. (Doc), 210–11
South Pueblo Ironworks, 121
Spanish expeditions, 45–47
Spanish Peaks Ranch, 157
Springs, 11, 148–49
Stanford Alumni Association, **138**
Stanford University, 30–32, 135–36
State National Bank, 39, 91–94, **93**
Stewart's Crossing, 176
Stone and Webster, 30
Swenson, Kristen, 222–24, **223**, 253,
    256–57
Swenson, Rock, 222–24, **223**, 253, 256–
    57

Tannery, Harold, 61–62
Taos pueblo, 24
Taos Trail, 28
Tapocott, Bill, 214
Terrell, Edwin R., 36, 241
Terrell, Shirley. *See* Leavell, Shirley
Terrell County, 36
Tewas, 25
Texas Homes Corporation, 62, 87, 91
Thailand, 125–29
Thomson, J. Edgar, 68

3R Ranch, 157
Three Rivers Ranch, 160–66
Titan missiles, 61, **61**, 143–46, **145**
Tiwas, 25
Trail to Disaster (Richmond), 56
Tubac, 28
Tubbs, "Poker" Alice, 102, **104**
Turner, Ted, 60, 205
Two–Thousand–Yard Stare (Lea), 214

Union Pacific Railroad, 112
Utah Construction Company, 62
Utes, 9–11, 23–26, 190

Van Horn Ranch, 158

Wagon Wheel Gap, 4, 6, **150**
Wagon Wheel Gap Ranch, 7, 80, 148.
    See also 4UR
Wagon Wheel Gap Springs, 191
Walters, Lieutenant, 176
Walton, Izaak, 261
Walton, Joe, 192, 253
Walton, Katus, 192–94, 253

Walton, William Wallace, 238–39
Warman, Cy, 109–10
Wasson, A. H., 109
West Moreland Coal Company, 68
White Sands Missile Range, 58–62, **59**
White, Byron, 206
Williams, Bill, 50, 52, 54, 55
Willow, Colorado. *See* Creede, Colorado
Wine Glass Ranch, 158
Winston, David R., 158, 168–70, 173,
    247–48
Winston, Fred, **243**, 248
Winston, Matthew, **243**, 248
Winston, Shelly, 243
Winters, T. W., 158
Wintz, Ed., 7, 191–92, 195–201
Wise, Bill, 232, **234**
Wolf Creek Pass, **43**
Woolley, Brian, 243
Wright, R. E. (Sonny), 166

Y.W.C.A., 85, 241–42
Yutas. *See* Utes